SUSTAINABLE
FOREST MANAGEMENT

SUSTAINABLE FOREST MANAGEMENT

Ian S. Ferguson

Melbourne

OXFORD UNIVERSITY PRESS

Oxford   Auckland   New York

OXFORD UNIVERSITY PRESS AUSTRALIA

Oxford   New York
Athens   Auckland   Bangkok   Bombay
Calcutta   Cape Town   Dar es Salaam   Delhi
Florence   Hong Kong   Istanbul   Karachi
Kuala Lumpur   Madras   Madrid   Melbourne
Mexico City   Nairobi   Paris   Singapore
Taipei   Tokyo   Toronto

and associated companies in
Berlin   Ibadan

OXFORD is a trade mark of Oxford University Press

National Library of Australia
Cataloguing-in-Publication data:

Ferguson, I. S. (Ian Stewart), 1935– .
   Sustainable forest management.

   Bibliography.
   Includes index.
   ISBN 0 19 5533604 2.

   1. Forest policy. 2. Forest management. 3. Forests and forestry —
   Economic aspects. I. Title.

333.7517

Edited by Jenny Missen
Printed by Australian Print Group
Published by Oxford University Press,
253 Normanby Road, South Melbourne, Australia

preface

This book is written for people interested in contemporary environmental management issues or pursuing studies in environmental science, natural resources, land management, or forest science. It represents an attempt to integrate knowledge and processes relating to forest management across a wide array of disciplines, but is centrally concerned with bringing the underlying biological and economic science together. The topic is not a simple one, notwithstanding the media treatment of it, nor are there panaceas to be dispensed. The management of publicly-owned native forests, the focal point of the book, encapsulates many of the most difficult problems in economics and biology. In an era when urban dwellers in general grow increasingly remote from experience of working in and understanding the forest and the complexities of the spatial and temporal scales involved, it is important to gain a broad perspective on these problems before attempting to solve them.

I owe thanks to many people. My parents encouraged me to pursue forestry following an interest expressed at an early age and it is a choice I have never regretted. My colleagues in forest services, especially the Western Australian service where I commenced my career, have been generous in sharing their knowledge and in debating the issues, as have my colleagues at the University of Melbourne and the Australian National University. In the international arena, managers of national parks and publicly-owned forests in many countries have gone out of their way to show me something of their work and problems. I have benefited greatly from formal and informal interaction with people in both the forest industries and the conservation movement. Jonathan Wearne and Sandra Roberts assisted greatly in the final preparation of the manuscript. My greatest debt, however, is to my wife Sandra, and daughters Claire and Heather, for being so tolerant of the peregrinations of a confirmed workaholic, not to mention the inevitable clutch of photographs of yet another region or country and its forests.

table of contents

figures

tables

1 introduction

The polar extremes of conservation and development of publicly-owned native forests are often vigorously argued by the respective interest groups, especially in the media. Biodiversity, climate change, protecting rare and endangered species and ecosystems, eco-tourism, water quality, protecting the interests of indigenous people, sustainable management, economic development, and ensuring community stability are but a few of the lines of argument that are advanced by one or other of the interest groups. International agreements are being pursued to achieve sustainable management of tropical forests, most of which are publicly owned, by the year 2000. Similar measures have been proposed for temperate forests.

Rational resolution of these controversies is not simple because many of the variables are difficult to define and measure, the time periods involved are long and clouded by uncertainties, and the institutional processes are complex. Furthermore, these controversies inevitably involve biological and economic factors, not to mention political, anthropological, legal and others with the added difficulty of quite different sets of terms, theories and principles.

The principal aim of this book is therefore to review critically and synthesise the underlying elements of these controversies and to place them within an ordered framework of processes for choice, not with any pretext of providing a universal solution, but rather of contributing to a better understanding and analysis of the issues. This may seem at odds with the title of a book that seems to promise a panacea in terms of sustainable management of public forests, yet it is the very essence of sustainable management. Those who seek or expect a single neat solution are bound to be disappointed. Those who recognise that

1

policy issues relating to publicly-owned resources are necessarily complex and varied in approach will have fewer problems.

PUBLIC FORESTS

Most, but certainly not all, major controversies over the use of forests are associated with those that are publicly owned, for the obvious reason that the forest manager, generally a government agency, is often beset with an array of apparently conflicting demands from people who perceive themselves as having an interest in that forest. Such perception sometimes extends well beyond the formal membership of the people in whose name ownership is vested, as witness the role of international and national conservation groups. Institutional processes for resolving these controversies are therefore seldom confined to one particular level of government.

Public ownership of forests is itself diverse in form. In some countries, public ownership is vested in the national government; in others, state, provincial or regional government; and in some in local government, tribal communities or clans. Governments of whatever level and form, other than totalitarian, are subject to influence from their constituents as well as from other levels of government and political action. Institutional processes for identifying property rights are thus central to resolving these controversies.

The notion of public ownership of forests is relatively new, although it has precedents in the 'commons' and similar types of communal ownership. In the sense of well-established and codified legislation, it evolved from the Crown lands owned by monarchs and royalty. As democratic governments evolved, the notion of Crown or public land became one of stewardship and holding in trust for the nation by the government concerned. More recently, the distinction between the 'Crown' (or government) as the legal owner and the public at large has become a source of discontent where that ownership has been pursued in a paternalistic manner, without seeking the views of the public or otherwise involving them in the decision-making process.

Why are these controversies so much more evident on public than on private land?

In general, private land lacks the qualities now often sought for contemporary conservation and recreation, and sometimes for water production, such as large areas of native old-growth or relatively undisturbed regrowth forest. The terms 'native' and 'old-growth' require some elaboration. Native (or indigenous) forests are those established by nature, or with comparatively little human intervention in the perception of the public at large. Comparatively little human intervention[1] generally means having or using regeneration from seed of the naturally occurring species of the site, either from seed trees, sown by hand or by plane, or planted by hand. Some restocking of poorly regenerated sites with planted seedlings may be tolerable, but broad-scale planting of nursery-raised stock is generally not tolerated, unless in the event of a calamity. Native species are sought because of the underlying desire to preserve and enjoy the natural environment, notwithstanding the considerable changes wrought upon it. The emphasis on native forest is also important because it largely excludes plantation estates[2] established publicly or privately for wood production, and certainly excludes those established with exotic species. Old-growth refers to forests whose age and structure are characteristic of a mature or over-mature ecosystem.[3] Older and relatively undisturbed regrowth forest arising from past clearing for agriculture or from long past harvesting is often difficult to distinguish from old-growth.[4] The issue of age is of lesser moment than of that naturalness or apparent naturalness.

The term 'public forest' is often used in this book as a contraction of 'publicly-owned native forest'. Public forests form the principal focus of conservation versus development issues.

Property rights are generally more clearly defined and identified for private land and thus owners normally have greater freedom to choose those actions that best suit their objectives. If compromises have to be made, they are made by the owner; either singly or through a board of directors in the case of a company or trust. If the public at large, acting through the government, wishes to impose changes on the exercise of property rights by the owner, such as by declaring a protected catchment or forest, the private landowner may be entitled to compensation.

In contrast, the political process makes publicly-owned land susceptible to claims by particular interest groups, generally without compensation where unpriced goods such as conservation or recreation are to be supplied. While there is nothing wrong or inappropriate in this, it places a greater emphasis on the need for careful evaluation and weighing of the choices.

Public ownership of native forests is widespread, as figure 1.1 shows. Data on ownership for Africa, Latin America and Asia are not yet available, but public ownership of forests in developing countries is the general rule. While defying precise geographic location and measurement of extent, those countries with considerable areas of publicly-owned and native old-growth or relatively undisturbed regrowth forest tend to be located around or within the Pacific Rim. Some countries, such as the Peoples' Republic of China, Chile and Mexico, have only small areas of such forest and the magnitudes of the controversies are correspondingly less. There are also countries, such as Russia, where major political changes are in train and the forest ownership issues have yet to be resolved, not to mention those where a near-totalitarian government currently prevails. The forest types involved are diverse, yet the major issues concerning conservation and development are common to most.

The appropriate analytical tools for resolving conflicts concerning public forests are those that involve choice between alternative uses. These lie in the domain of economics as far as choice is concerned, but centrally in the domain of biology where production possibilities involving alternative uses are concerned. Thus an introduction to the economics of choice between alternative uses is relevant here because the terminology and analytical tools form important precursors for subsequent chapters. Of course, neither economics nor biology provides a complete theory or set of tools for the purpose, and other areas will be introduced as appropriate throughout this book.

THE SOCIAL NET BENEFIT CRITERION

The underlying criterion used in economics is simple. One should choose that land use or combination of uses that provides the greatest social net benefit to the society concerned. In principle, this is clear and unobjectionable, provided one accepts an anthropocentric orientation. In practice, it raises almost as many questions as it answers. Many books have been written on the analysis of social net benefits.[5] Some, such as McNeely's book[6] on economics and biological

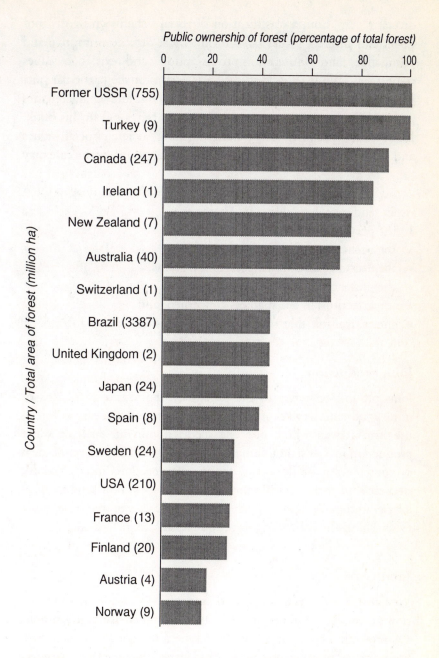

Figure 1.1 Public ownership of forest in selected countries. (Data of Food and Agriculture Organization and United Nations-Economic Community of Europe 1993, p.113, collected between 1980 and 1990; and of Arthur Anderson Co. in Cubbage et al. 1993, Fig. 18-4).

diversity, start from a classification of social benefits and costs into direct and indirect values, and within those into consumptive and productive, and non-consumptive, option, and existence values respectively. While this classification is useful for the particular purpose, it is ill-suited to reviewing the economic characteristics involved in the broader range of forest uses canvassed in this book. Many of the problems lie in clarifying how 'society', 'social' values and 'net benefits' can be defined and analysed; or not, as the case may be.

Six characteristics associated with conservation and development issues render analysis difficult. These are:
- the pervasiveness of joint production,
- the occurrence of joint costs,
- the existence of external costs and benefits,
- the presence of risk and uncertainty,
- imperfections in the markets concerned, and
- the critical role that time plays in most aspects of production.

Joint production

Joint production refers to the potential for production of two or more goods and services at a time, as is commonly the case in forest-use issues. Even where one use seems dominant, such as water production in a closed catchment, there are generally other uses, such as conservation of flora and fauna, that need to be recognised. Because some of the terminology associated with joint production is central to describing relationships between uses, we shall have more to say about the theory of joint production later in this chapter.

Joint costs

Joint costs refer to the difficulty of attributing inputs to the production processes unambiguously to one or another use. For example, fire protection costs for a closed catchment cannot always be attributed entirely to water production and, as we shall see later, there are particular complications if fuel reduction burning is used as part of the protection process. Although difficult, the solution to the allocation of joint costs is normally to look first for the motivation behind the use of the input—most if not all of the cost should normally be carried by the use that is driving it. Where a clear cost driver is not evident, then a subjective judgment of the proportional allocation must be made by those associated with managing the area. If no basis can be found for a subjective allocation, it may be necessary to treat the costs concerned as general overhead costs, but modern accounting practice is to allocate wherever possible.

External costs and benefits

External costs and benefits refer to those costs and benefits that are borne by or accrue to individuals other than the owner of the forest. Examples include possible damage to downstream water quality from harvesting operations in forests, or access to recreational areas provided by roads constructed for harvesting. Externalities are pervasive in conservation and development issues but are often difficult to value. Data on the physical outcomes are often lacking or expensive to collect and prices may be lacking because of market imperfections.

Risk and uncertainty

The issue of data collection is especially vexing because some of the principal external effects are associated with extreme events, such as high intensity precipitation, drought, cyclonic or very high winds, or wildfire. Not only is it difficult and expensive to measure the effects of these events on the use concerned but it is more difficult to model future occurrences. Extreme events highlight the presence of risk and uncertainty, although both are more pervasive than these events alone.

Risk defines a situation in which the probabilities of the different outcomes occurring can be gauged by reference to past records or other objective measures, while uncertainty arises where the probabilities of the different outcomes occurring cannot be gauged other than subjectively. The distinction is not of major consequence here. However, recognising the stochastic nature of any set of outcomes is important, especially in matters of population viability, where the probability of a population falling below a critical level is a measure of impact. In an ideal world, the probabilities attached to all sets of benefits and costs should be taken into account. In practice, all that can be done may be sensitivity analysis of the final solution to changes in the components.

Market imperfections

Imperfections in markets are not confined to conservation and development issues, but they are notable because many forest uses do not have the property of excludability that characterises an effective market process.

Passive conservation, in the sense of knowing that particular species or ecosystems are preserved for future generations, is a use enjoyed by many; there is no way of excluding a person from that enjoyment and hence it would be difficult to develop a market and charge for that use, if one was so inclined. Another way of expressing this is that passive conservation is a public consumption good,[7] in that the amount consumed by one person has no effect on the amount available for others to consume.

Even where excludability is not a problem, as in the case of timber sales, the spatial distribution of forests relative to centres of timber consumption may lead to imperfections in market operations such that prices paid do not properly reflect social values. Four approaches are used to provide prices that more aptly reflect social values.

One is for commercially tradeable[8] commodities, such as timber, where the c.i.f (cost including freight) price of imports or the f.o.b. (free on board) price of exports may be used as a shadow price with appropriate adjustment for domestic transportation and handling costs. Where foreign exchange rates are set artificially high to encourage exports and discourage imports, shadow exchange rates also have to be applied to correct for these distortions.[9]

The second is for commercial inputs, such as labour and capital, where the imperfections, if present, cannot be gauged directly from internationally traded values. In these cases, more complex calculations are needed to evaluate 'shadow' prices because of the macroeconomic relationships involved.[10] Of course, such shadow prices are only needed where marked distortions in markets occur and these tend to be more typical of developing countries. Examples are where wage rates for labour are set at artificially high levels to encourage the movement of labour from a rural subsistence economy to the cash economy; or where capital markets are distorted by substantial government investment but government enterprises are free from company taxes and other imposts.

The third is for non-tradeable but commercial inputs, including water, electricity and gas. Here the process is to disaggregate into capital and labour categories and value accordingly.

The fourth is for non-tradeable non-commercial outputs such as conservation or dispersed forest recreation, for which no prices exist. Here, shadow prices have to be imputed from indirect evidence such as contingent valuation,[11] travel costs[12] or related techniques.[13]

Time

Finally, time is a dominant element in the production process for many forest uses. Moreover, the time scale frequently extends well beyond the horizons common to manufacturing industry and commercial and financial markets. A thirty-year rotation is a relatively short period of production in most native forests; many could extend to 200 years, especially for conservation. This means that the processes of discounting for time necessarily become interwoven with issues of equity between successive generations, the subject of a later chapter.

Of the six characteristics enumerated above, only joint production needs immediate elaboration, because it is central to the subsequent description of the relationships between uses.

JOINT PRODUCTION

The economic theory of joint production[14] deals with any problem involving the potential production of more than one use or service from a particular area of land, whether the production of those uses be mutually exclusive or not. Let us assume initially that they are not mutually exclusive, as for example wood production and dispersed recreation over a sizeable tract of forest. The properties of each of these uses will be developed in detail in later chapters. They are used here as generic examples.

There are two separate components to the theory of joint production; one belonging to the biological domain of the function that reflects the set of production possibilities (or inter-relationships) between the two (or more) uses and the other that reflects the economic domain of the values or prices to be applied to each.

Production possibilities

At one extreme, if harvesting is proceeding over the entire area, the scope for dispersed recreation is virtually zero. On the other hand, if recreation is proceeding over the entire area, the scope for harvesting is similarly zero. In between, there are a set of combinations of those uses that reflect the feasible production possibilities. Some possible sets are shown in figure 1.2.

The sets shown in figure 1.2 have been chosen to illustrate a range of quite different relationships, according to the trend of the slope (known as the marginal rate of transformation) between the two uses. Each set assumes that the value of the inputs used in providing any one combination is constant.

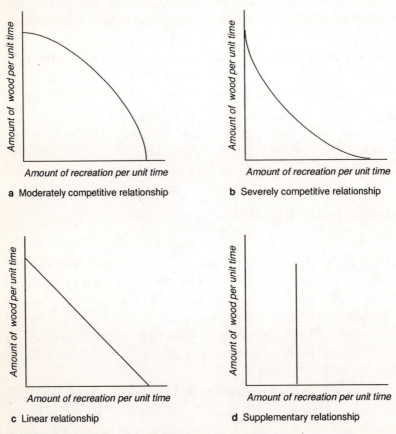

Figure 1.2 Sets of production possibilities.

If the trend of the slope is declining, for each additional unit of recreation pursued, proportionately more recreation is being added than the proportion of wood foregone and the relationship is called moderately competitive. If the trend of the slope is increasing, the converse is true and the relationship is called severely competitive.

The other two cases are less likely or are limited to a relatively small range of production possibilities, but are included for completeness. The first is where the slope is constant but neither zero nor infinite and is called a linear relationship.[15] The second is where the slope is constant and either zero or infinite (only the latter is illustrated) and is called a supplementary relationship.[16]

In practice, the shapes assumed may vary along the surface. A set does not have to be entirely complementary or entirely competitive: it may change from one to the other along the surface and have supplementary or linear regions as well. It may also change with increasing forest age or with forms of treatment. Nevertheless, the capacity to describe the surface, even if only qualitatively by reference to these shapes, can be a powerful aid in indicating the best combination of uses.

Prices

Let us assume that there is but a single owner of the particular tract of forest under consideration and, further, that his or her objective is simply to maximise revenue. Let us further assume that the owner is operating in a world of perfectly competitive markets; implying freedom of entry and exit, many small producers and consumers, and perfect information. At a later stage, we shall consider deviations from these assumptions, but initially they are convenient simplifications.

Under these assumptions, the prices received for these goods or services reflect their social as well as their private values and are invariant to changes in the amounts that the owner might choose to place on the markets concerned. The revenue received will be the sum of the product of the price and the amount of each good or service supplied. Iso-revenue lines can be represented on figure 1.3 by straight lines whose slope is equal to the negative of the ratio of the prices. The highest revenue achievable is represented by the iso-revenue line that is furthest to the right and just tangent to the curve representing the set of production possibilities as illustrated in figures 1.3a and 1.3b. The others cannot achieve tangents, being limited by boundary conditions, but the principle is the same, as shown in figures 1.3c and 1.3d.

Of course, any one of these sets of production possibilities assumes a given constant level of investment in the provision of inputs. If a higher level of investment were available, the set would shift outwards from the origin and establish a new frontier and optimum combination according to the same principle. For the time being, however, the simple categorisation of shapes of the set of production possibilities will suffice.

Note that nothing in the theory dictates that the optimum combination should be a mixture of uses—multiple use in the parlance

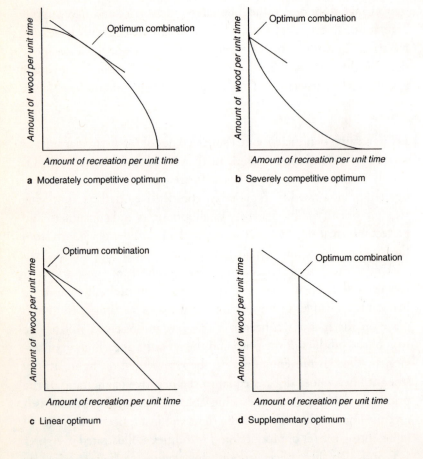

Figure 1.3 Optimum combinations of uses.

of forestry.[17] If the relationship is severely competitive, and generally when it is linear, then exclusive use of one or the other will be optimum, depending on the relative prices. If the relationship is supplementary or complementary, a combination will be optimum. If the relationship is moderately competitive then a combination may be optimum, provided both prices are positive. If one or the other price is zero, then a single use becomes optimum.

SCOPE OF THE BOOK

Part I of the book deals individually with the properties of some of the major forest uses or services, so that a clear perspective may be gained of the characteristics of production. In Part II, the resolution of conflicts between uses and services is pursued at various levels of decision-making. Part III deals with some outstanding issues and summarises the conclusions.

forest uses

There are many uses of forests; too many to be described here. Instead, the focus in Part 1 is on the principal or most common uses, whose properties cover such a wide range as to provide useful examples of the principles involved in almost all other uses. One of the difficulties in describing such a range of uses and properties is that it requires reference to technical terms drawn from an array of disciplines. But that simply highlights the multi-disciplinary character of forests and forestry and the challenges inherent in their study.

Part 1 deals first with conservation as a forest use: in many ways the most difficult area because of the relatively recent evolution of views about the various properties ascribed to it. Chapter 3 deals with recreation, a use that is diverse in character but largely directly consumptive in nature. Water, which is taken up in chapter 4, is also directly consumed by the final user, but the intervening storage and reticulation measures tend to obscure its economic characteristics at the forested catchment. Finally, chapter 5 deals with wood, a diverse product with many uses that are variously directly and indirectly consumptive and durable and non-durable.

2 conservation

CONSERVATION AS A FOREST USE

Conservation is a word used in many different ways. In this study, it is defined as the maintenance of plants, animals and ecosystems in an essentially natural form that enable natural processes to continue. It is therefore appropriate to treat it as one of a number of possible uses of forests; a use that is separate and distinct from recreation, hydrology, wood production and other uses of forests.

Forest conservation, in this sense, provides benefits in a number of different ways such as:
- protecting other living things,
- improving the environment, and
- contributing to material needs.

Protecting living things

Conserving native public forests in order to protect living things rests in part on a willingness to recognise that species have a moral right to exist and thus that areas containing threatened species may need to be reserved. The United Nations General Assembly acknowledged this moral right in the World Charter for Nature in 1982.

Ensuring the persistence of a species does not imply that all individuals should necessarily be protected but rather that the species should not be allowed to become extinct as a result of human intervention. Mention of human intervention is important because some species will become extinct naturally, just as new ones will evolve. Of course, some people hold the view that any living thing has a right to life, some that only animals do, and some that the issue is one of avoiding cruelty rather than of preserving life.[1] In a democracy, the government has the responsibility of reflecting the views of the majority on such an issue and almost all have chosen the latter position, not the former positions.

Part of the motivation for protecting species may also be a desire to endow heirs and contemporaries with the opportunity to see them, and perhaps to conduct research on them, and has a more utilitarian than moral basis. This is known as custodial responsibility in the conservation literature. Regardless of the form of motivation, there remains the question of how such motivations and values can be interpreted in economic terms.

Goods and services that provide satisfaction to people through the knowledge that species (or ecosystems) exist and will continue to be protected, even though they have no intention of using the species or visiting the area are said to have existence values[2]. These goods and services are generally public consumption goods because consumption of that knowledge by one person has no impact on the ability of others to consume it. Nevertheless, there have been many occasions when groups of citizens have acted privately to protect a species or area. For example, when one of the last three habitats of the threatened hairy-nosed wombat in South Australia was threatened with development for agriculture in 1967, a private trust was formed to raise funds and successfully purchased the area[3] at current market value. The donors thus endowed their contemporaries and heirs with a public consumption good. The distinction between public and private conservation is therefore not in the outcome but in who bears the cost of creating or supporting the provision of a public consumption good. In most cases, the taxpayers at large bear the costs of protecting species and areas through the creation of reserves in publicly-owned native forest.

Conservation biology has undergone considerable development over the past thirty years. The term 'threatened species' reflects the early history of conservation in the 1960s when attempts began to document the depletion and possible extinction of species. Lists of globally threatened species were drawn up, using categories defined by the International Union for Conservation of Nature and Natural Resources. In abbreviated form, the definitions[4] that seem likely to be adopted in the near future are as follows:

Extinct: a taxon is extinct when there is no reasonable doubt that its last individual has died.

Extinct in the Wild: a taxon is extinct in the wild when it is known only to survive in cultivation, in captivity, or as a naturalised population (or populations) well outside the past range.

Critically Endangered: a taxon is critically endangered when it is facing an extremely high risk of extinction in the wild in the near future, as indicated by a quantitative analysis showing the probability of extinction in the wild is at least 50 per cent within five years or two generations, whichever is the longer.

Endangered: a taxon is endangered when it is not Critically Endangered but is facing a very high risk of extinction in the near future, as indicated by a quantitative analysis showing the probability of extinction in the wild is at least 20 per cent within twenty years or five generations, whichever is the longer.

Vulnerable: a taxon is vulnerable when it is not Critically Endangered or Endangered but is facing a high risk of extinction in the wild in the medium term future, as indicated by a quantitative analysis showing a probability of extinction in the wild of at least 10 per cent within 100 years.

Each of the latter three categories has alternative criteria based on population sizes, percentage decline in numbers, or extent and fragility of occurrence that can be applied in the event that the data for a quantitative analysis of risk are insufficient or the risk model is inadequate. In addition to those listed, there are categories of Conservation Dependent, Low Risk, Data Deficient and Not Evaluated, which are largely self evident in definition.

The notion of conserving species or areas is now seen as having broader dimensions than that of conserving threatened species. Increasingly, emphasis has been placed on conserving biological diversity in three inter-related forms:

- genetic diversity,
- species diversity, and
- ecosystem diversity.

Genetic diversity

In an era when genes can be transferred readily from one organism to another by genetic engineering and deliberate breeding strategies, there is a risk of losing a collection of potentially useful genes in that process. Naturally occurring species provide a stock of those genes. As experience in breeding commercial crops such as wheat, maize[5], rice and plantation-grown pines has shown, it is often important to have access to wild species in order to develop a commercial species for new or changing conditions or improved or different properties. Conservation and other reserves and other forms of gene banks provide a means of maintaining a wide representation of the naturally evolving forms. Conservation reserves, as distinct from other forms of representation, also provide a medium in which natural evolutionary processes can continue.

It can be argued that natural evolution is unattainable because all remaining forests are influenced to some degree by recent human disturbance, if only through global warming. For example, according to current forecasts, to keep pace with global warming over the next 100 years, temperate-zone species with narrow latitudinal ranges may have to migrate towards the relevant pole at a rate of 3000 m per year, some ten times faster than the migration that followed the retreat of the last glaciers.[6]

Even so, there remains an argument for conserving areas where evolutionary and migratory processes can proceed with minimal direct human interference. Operationally, however, conserving genetic diversity lacks any specific spatial definition. In a practical sense, it is species (or provenances of them) and ecosystems that have a ready spatial identity. Thus genetic diversity is more a statement of good intent than a useful objective for determining conservation uses.

Species diversity

The argument for species diversity is an extension of the threatened species argument. In order to ensure survival of a particular species, it may be necessary to focus attention on its habitat, rather than the species alone. Maintaining the natural diversity of the habitat may be an important component in this process and may require specific measures to protect against invasion by exotic plants and feral animals.

Relatively simple relationships between the number of species and the area of a reserve were initially said to govern species extinction:

- S = ρA^z —the Arrenhius relationship
  where S denotes the number of species,
- ρ is a constant,
- A is the area of the reserve, and
- z is a power coefficient varying between 0.15 and 0.40.

The widely quoted generalisation that a loss of 90 per cent of habitat will result in a loss of between 30 per cent (z=0.15) and 60 per cent (z=0.40) of species is derived from this relationship. So too, if interpolated more crudely, is the statement that a ten-fold decrease in area leads to a loss of half the species. However, the relationship is really an artefact of the combined effects of sampling and species heterogeneity and tells us little about the specific probabilities of species extinction.

This very general theory of population dynamics was to be super-seded by the theory of island biogeography;[7] scattered conservation reserves being analogous to islands in terms of the capacity of the populations of species they contain to survive. The insights provided by this theory are now generally considered to be an inadequate basis for detailed recommendations on conservation because they ignore species identity, habitat heterogeneity and population sizes. They have such wide margins of error that they have low explanatory power and give unreliable estimates.[8] Nevertheless, the principles involved in the theory have given rise to some useful guidelines about the design and management of reserves and their surrounding areas.[9]

Species become extinct for a variety of reasons. These include direct cause and effect relationships such as human intervention or glaciation, and many indirect chance events affecting genetic consti-tution, the survival and reproduction of individuals or environmental changes and catastrophes.

Extinction is best analysed through specific models of the popu-lation dynamics of a target species based on its ecological responses, thereby providing information about population size and the proba-bilities of extinction. Risk categories can be assigned according to the probabilities of extinction or survival,[10] thereby providing a more objective system for categorising threatened species. However, these models demand considerable data and it will be some time before all the most obviously threatened species can be examined.

CASE STUDY 2.1 Habitat management and protection for Leadbeater's possum.[11]

Leadbeater's possum (*Gymnobelideus leadbeateri*) is an arboreal mammal found only in certain areas of the Central Highlands of Victoria, Australia. Adults are small, weighing 100–160 g, with a distinctive club-shaped tail as shown in figure 2.1.

Prior to 1961, Leadbeater's possum was presumed extinct. Since rediscovery, controversy has existed regarding the appropriate management and protection of the forests that sustain the remaining populations, these forests being of high commercial value for mountain ash timber.

Surveys indicated that the presence of Leadbeater's possum is significantly associated with the abundance of overstorey trees with nesting hollows. The age of the mountain ash trees with nesting hollows is typically at least 190 years and may exceed 400 years. The presence of the possum is also significantly associated with the density of understorey trees of *Acacia* species that provide edible exudates of gums, some of which appear to be

Figure 2.1 Leadbeater's possum

rich in sugars and proteins. However, the bulk of the dietary protein for the species is obtained from insects associated with the strips of bark that hang from the trunk and branches of the mountain ash trees. Surveys also indicate that linear strips of old-growth or mature forest, retained after harvesting eight years earlier and ranging from 30–260 m in width, seldom contain the possum because of the dispersal of food sources. Thus the amount and spatial configuration of suitable habitat on a catchment or landscape scale may be important in determining the distribution and abundance of the possum.

A mathematical model was developed to enable population viability analysis to be carried out. Using data on life cycle attributes, habitat requirements and spatial distribution, temporal changes in habitat suitability, aspects of timber harvesting, and the frequency and severity of wildfires, various simulations were carried out on four different blocks of forest—one totally protected, and three with timber harvesting permitted, including one with a very small and fragmented area of residual old-growth forest.

Drawing on the results of these simulations, guidelines were developed for a reserve system within those blocks in which timber production is to be permitted. These guidelines include reserving all large (over 10 ha) patches of residual old-growth forest within or near to a series of twelve to twenty reserved areas, each of 50–100 ha. This multiple-patch network provides a risk-spreading strategy designed to minimise the chance that all reserved areas would be destroyed in a single wildfire. Special care needs to be taken to reduce the risks of reserved patches being burnt in the course of regeneration burns. In the event of a wildfire, no salvage harvesting of these patches should be permitted.

In coupes which were harvested, all possums were assumed to die following harvesting and regeneration burning. Fieldwork and observation[12] in the mid-1950s suggest that this assumption is incorrect. Many of the very patches studied in one block harvested then now contain colonies. The only patches of old-growth left at that time were seed trees, mainly trees with hollows or other defects rendering them unsuitable for timber. Apart from some very occasional small clumps of trees in areas too difficult to log, no patches or clumps of trees were deliberately left, although the random distribution of trees with defects could give rise to that appearance much later in time. While recolonisation from uncut areas may have been responsible for restocking, this seems unlikely in view of the extremely severe wildfire that burnt the surrounding areas in 1939 and the severity and rapidity of harvesting of the remnant old-growth in the 1950s.

The death of all possums following harvesting and regeneration burning is a critical assumption that deserves further study, although this is not to argue that there were no deaths. Despite these reservations, the approach to modelling is appropriate and the general thrust of the guidelines is likely to remain valid.

In general, models of the effects of chance events suggest[13] that:

- demographic chance events are only a hazard for relatively small populations numbering tens or hundreds;
- no population size can guarantee a high level of long-term security against environmental conditions and catastrophes;
- increases in population size yield diminishing returns in persistence times for a given disturbance; and
- large population sizes or numerous populations are required in order to be reasonably certain of conserving a species for a significant length of time.

However, many qualifications apply to these generalisations. For example, small populations that occasionally undergo moderate levels of stress may be more appropriate for conservation than similar-size populations in benign environments.[14] Given the relatively long life span, persistent seed pools or vegetative propagation of many plants, small reserves spanning a wider choice of sites may play an important role in plant conservation—more so than for animals.[15]

Much depends on the mobility and range of the species. A population of herbaceous plants of very restricted distribution may be maintained quite effectively by delineating and protecting a local conservation reserve. An animal such as an owl, with a wide range but demanding habitat for nesting, may require conservation measures to be adopted very widely on both conservation reserves and intervening areas. Notwithstanding the controversy that can surround the economic implications of the latter cases, the technical development of conservation reserves and related measures for threatened individual species is fairly straightforward once sufficient information is available to model the dynamics of the population/s concerned. In the absence of this information, we can only resort to guidelines derived from similar or 'indicator' species.

Ecosystem diversity

Consideration of diversity within and between habitats and the need to reserve and protect a sufficient array of representative forest ecosystems, communities, alliances or types—depending on the scale and terminology one wishes to pursue—leads to consideration of biological diversity on an ecosystem, and ultimately on a regional, basis.

Models of ecosystems involve far greater complexity and data than is generally the case for threatened species. In particular, the structural, spatial and temporal interactions and effects become more important. Obtaining data that span all aspects is often difficult and hence the number of well-developed models of ecosystems is small. The 'Jabowa' model[16] and others that followed it[17] are examples. However, they are principally used to predict the outcomes of catastrophes and human interventions rather than as a basis for reservation itself. With few exceptions, modelling populations of many species in a region on a spatial basis is beyond our capacity, partly for want of the necessary data and partly because of the enormous scale of the model required. Reservation is more the province of descriptive analyses of regional vegetation types and assessment of the adequacy of their representation in the reservation system.

Regional analyses are generally undertaken by sampling the vegetation in the field, often following identification of relatively uniform forest communities or types from aerial photography,[18] or remote sensing based on tree canopy characteristics. These data are then related to environmental classes defined from computer-based mapping[19] of soils, topography and climate, to see whether the proposed reserve system has an adequate representation of communities or types across environmental classes.[20] Where possible, the reserve system should represent the heterogeneity of the community or type across its environmental range, not just a sample of one part of it. In other studies, where effective forms of stratification have not been available, vegetation types have been gauged by analysing floristic and/or physiognomic data from systematic grid surveys,[21] but an otherwise similar approach has been used in attempting to assess the adequacy of representation.

Operationally, conserving diversity leads to the identification of quite arbitrarily defined 'representative' reserves of widely varying size: some dedicated to the conservation of a particular threatened species; some to sets of species; and others to particular forest ecosystems, alliances, types or combinations thereof. Given a desire to reserve a particular species, ecosystem or sets thereof, the delineation of most conservation reserves is generally determined oppo tunistically by what relatively undisturbed areas are left and the alternative uses for these. Not surprisingly, controversies over old-growth forests have intensified as the remaining areas have diminished, because these forests have generally been least disturbed by humans and are often highly valued for wood production.

A minimum level of 5 per cent of the original land area of each major type has been recommended as a target for conservation reserves, although a threshold of 10 per cent has recently been recommended for the main ecological regions of a country.[22] However, these are very rough guides and lack any necessary or scientific relationship to the objective of providing an adequate system of conservation reserves. The delineation of conservation reserves merits more specific consideration than reliance on a particular proportion of the area to be reserved. Reserving 90 per cent of a particular type may be insufficient in some cases and 5 per cent more than sufficient in others.[23]

Such crude prescriptions for the extent of conservation reserves entirely ignore the critical objective, which is to provide a comprehensive system for ensuring that the risk of loss of particular forest types, ecosystems and species is kept to an acceptable level. In many, if not most, cases that risk will depend not only on the formal system of conservation reserves but also on the extent, distribution, degree and frequency of disturbance in other public or private forests. What constitutes an acceptable level of risk will also vary according to the species concerned. To take an extreme example, a rare grass or single tree species ecosystem may be relatively easy to re-establish from seed in seed banks, such that a higher than normal level of risk may be quite acceptable. A species whose seed deteriorates in storage and is sparse and irregular in the quantity of seed set may require setting a lower level of risk.

Much more attention ought to be paid to defining what is an acceptable level of risk, because the assertion of absolutes is neither attainable nor practicable. For example, species richness is often used as a criterion for conservation in the name of biological diversity because it often reflects stable evolutionary processes relatively uninterrupted by episodes of extinction during unfavourable climatic periods. The so-called Pleistocene rainforest refugia of Africa and

America[24] are especially important examples. However, species richness is an ambiguous criterion. Many lowland rainforest communities in Malesia are rich in plant and animal species that coexist as a result of historical chance from the ebb and flow of climatic change; not necessarily because those species coevolved. As Whitmore[25] points out, these communities are 'neither immutable nor finely tuned'. This highlights the importance of using the risk of loss of the type, community, or ecosystem due to human intervention as the ultimate criterion, rather than stability or richness.

Improving the environment

Forests may contribute to improving the environment in several ways; the capacity of forests to reduce atmospheric concentrations of carbon dioxide through photosynthesis being the best known.

The widely promulgated forecasts of progressive global warming associated with increasing concentrations of carbon dioxide, nitrous oxide, methane and chloro-fluorocarbons in the atmosphere have prompted extensive analyses of the potential role of forests in ameliorating this problem. A study by the World Meteorological Organisation and the United Nations Environment Programme[26] has advocated several steps including:

- reducing the burning of fossil fuels to reduce the present release of about 5.6 billion tonnes of carbon annually;
- reducing deforestation, which currently accounts for release of some 1 to 3 billion tonnes of carbon annually;
- reforesting 100 to 200 million ha to absorb about 1 million tonnes of carbon annually.[27]

The reforestation figures, if implemented over a fifty-year period, imply an enormous expansion in the rate of reforestation[28] and a very large investment of resources. Acceptance of these recommendations is unclear at this stage because of the uncertainties attached to many of the key parameters[29] used in the forecasts and because the changes are distributed unevenly across the globe; with some regions gaining, others losing.

Sulphur dioxide is adsorbed by macroscopic plants; carbon monoxide concentrations are reduced and oxidised by soil fungi and bacteria; and nitrous oxide is incorporated into the biological nitrogen cycle.[30] Notwithstanding these potential benefits, no cases are known where the conservation of native public forests has a specific relationship to them. It seems likely that any such cases would be very site specific and limited in extent.

Long-standing assertions that forests increase rainfall are difficult to prove and are generally discounted relative to the effects of other forms of vegetation. There are, of course, localised effects, such as those due to the additional height of the forest and the roughness of the canopy, that may increase the effective precipitation from fog or cloud drip in montane forests[31] and provide a basis for conserving the forests concerned. The Amazon Basin is another localised, but very extensive, case in which the inland appears to receive about half of its rainfall from recycled water. The moisture-bearing Atlantic trade winds maintain the rainforests in the lower Basin. These return a high proportion of the moisture to the atmosphere through evapo-transpiration and this is carried towards the Andes and falls again as inland rain. Removing the rainforest would reduce the evapo-transpiration and increase river flows at the cost of inland rainfall,[32] again providing an argument for conservation or at least for very conservative management.

Contributing to material needs

Forests provide a variety of material needs. In addition to the major products requiring more intensive use of the forest, many other products are or could be derived extensively from forests with a min-imum of disturbance and hence interference to the natural pro-cesses. These products include edible fungi, leaves, roots, fruits, nuts or exudates; edible animals and fish; and pharmaceuticals, oils, fats, waxes and resins.

The role of forests in the supply of edible products is well known. Many edible and utilitarian 'minor' products derived from the forest are raised commercially by intensive means outside public forests because such production is cheaper and more readily controllable in terms of quality and protection.[33] Honey is an obvious exception in temperate hardwood forests. Other notable exceptions in developing countries are where forest-dwelling tribes and adjacent populations may depend substantially on forest sources of these products for their subsistence. There is evidence that edibles can play an important role in supporting local communities and that their value for this purpose may well exceed that derived from harvesting wood. A study by Ruitenbeek[34] indicated that 'minor' forest products represented some 29 per cent of income for the Korup region in the Cameroons, compared to 22 per cent for hunting and trapping and some 34 per cent for cash crops. The present values of net benefits attributable respectively to subsistence production largely based on these edibles in the Korup forest was equivalent to US$1.7 million, compared with US$1.2 million derived from harvesting wood. In the support zone around the forest, a further present value of US$7.0 million was generated, mainly from hunting and trapping.

While the role of native forests in providing oils, fats, waxes and resins has become less prominent as plantations, synthetics and petroleum-based derivatives have taken their place, there has been a recent resurgence of interest in potential pharmaceutical compounds in tropical forests. Some three-quarters of the world's population, mostly in developing countries, are said[35] to depend on plants as sources of medicines. Some 25 per cent of the prescriptions dispensed in the USA between 1959 and 1973 contained active ingredients derived from vascular plants.[36] Some of the major drugs in current use such as aspirin, oral contraceptives, cortisone and quinine were originally discovered as or derived from compounds in plants.

Given the enormous diversity of plant genera and species, there are undoubtedly many more useful compounds yet to be discovered. However, it does not follow that this untold store of useful compounds is automatically of great net benefit and thus an incontrovertible basis for conserving forests. It that were the case, one would expect that the very large pharmaceutical companies would long ago have staked their claim by outbidding the timber companies in order to conserve these forests. Recently[37] some pharmaceutical companies have taken out contracts for exclusive access to screen tropical rainforest specimens collected by biological survey agencies. It remains to be seen whether this will make the biological survey profitable or simply offset some of the costs otherwise borne from public sources. Experience with an intensive program of phyto-chemical surveys in Australia over the period 1920 to 1950 was disappointing,[38] but new analytical techniques and knowledge of key compounds may offer greater hope of success. Without detracting from the possible merits of this forest use relative to others, the general consensus of current opinion is that while some exciting and useful compounds may be discovered, they will not provide a sufficient basis for ensuring broad-scale conservation.

From the viewpoint of a potential investor, the situation is not dissimilar to mineral development. There is a high probability that most compounds or materials yet to be discovered will be of little social value and a low probability that some, quite a small number, will be of high social value. Unlike the case with minerals, however, it is difficult to exclude other companies from obtaining the species in question once a useful plant is discovered. For this reason, unless guided by specific information from traditional medicines and the like, few pharmaceutical companies are willing to engage directly in or to pay fees for field prospecting. Field survey is left principally to public or quasi-public agencies, who may then make their collections available to private companies to screen and assess the commercial prospects.

Conserving native forest is not the only means of maintaining access to these compounds. Arboreta, botanical gardens, seed banks and gene banks also serve this purpose, but by definition are limited to the species that are known or being identified. Given the considerable costs involved in these forms of preservation, they are probably best seen as an insurance mechanism against natural risks to families and genera of known or possible importance, and as an experimental and educational resource.

Finally, the commercial collection of materials for pharmaceuticals and traditional medicines sometimes involves destructive methods similar to timber harvesting. This is unlikely to be compatible with other motives for conservation of the area if carried out on any substantial scale.

CASE STUDY 2.2 The supply of traditional medicines from tropical rainforest in Belize, Central America.

Tropical rainforests in Belize, Central America, provide a range of goods and services, including roots, tubers, bark and fruits for consumption, domestic use, ceremonial or art purposes, and traditional medicines, to the people who live in close association with the forest. Traditional medicines play an important role in health care in Belize. According to local estimates, up to 75 per cent of primary health care is provided by way of traditional medicines and therapies.[39]

Preliminary destructive studies in two stands of different ages in Belize indicated that the present value of the net revenues to forest dwellers collecting and supplying the raw materials (variously bark, leaves, fruits, roots or tubers) ranged from US$726 to 3327/ha; figures that may exceed the values for the use of the same sites for wood production or intensive agriculture.[40]

An interesting approach to conservation is being tried at the Terra Nova Rainforest Reserve in Belize.[41] Among other things, the project aims to develop policies on intellectual property rights concerning the commercial use of ethno-botanical knowledge, to conduct research on the sustainable harvesting and extraction of the economically important components for traditional medicine, and to encourage eco-tourism.

Traditional methods of collection use both destructive and non-destructive techniques for harvesting or extracting the desired product from the vegetation. Potentially destructive techniques, such as bark stripping or removing roots or tubers, may be moderated by partial bark stripping to avoid completely girdling the tree, restricting the extent of root or tuber disturbance, or limiting felling or leaf or fruit harvesting to small patches at a time, together with providing an adequate period of time for the vegetation to recover or regenerate. The period for recovery or regeneration may be fifty years or more.[42] While the precise manner of sustainable harvesting and extraction and associated investment in ameliorative measures are still under research, many of the principles are similar to those for the sustainable timber management. However, the issues relating to this form of forest use are not solely those concerning the manner of harvesting. Much rests on developing formal recognition of the intellectual property rights, and effective markets in which the collectors are not exploited. Also, reliance on only a few medicinal products is fraught with risks relating to substitution and cyclical changes in the markets. The consensus of expert opinion[43] is that diversification of non-timber uses (i.e. joint production) will be required if sustainable and economically viable forms of forest use are to be developed for dependent forest dwellers. The relationships between the various possible non-timber uses (and between these and timber) have been studied very little, if at all, and need elucidation if efficient, sustainable combinations of production are to be identified.[44]

BENEFITS AND COSTS OF CONSERVATION

In general, the goods and services involved in contributing most of our material needs are market goods with established prices. In a few cases, a price may not be charged because of the costs incurred in extracting payment and a price has to be imputed for the purposes of analysing alternative uses. In others, the price charged may be distorted by some imperfection in the market, such as the exercise of monopoly power or tariff or non-tariff distortions of trade. In such cases, shadow prices may have to be derived from the actual prices, using adjustments to remove the effects of the market imperfections.

The benefits from protecting living things are not traded in the market because they are by nature public consumption goods. In those very few cases where a group of citizens is sufficiently concerned to raise funds privately and purchase an area, we have a tangible expression of a willingness to pay for the benefits anticipated. The 'willingness to pay' concept is capable of extension to society at large as a measure of the economic benefits associated with non-market or unpriced goods and services such as conservation. Willingness to pay is defined as the amount people would be willing to pay rather than forego their consumption. Consumption in this case is associated with the existence value of the area—the knowledge that the species or area concerned will be maintained for future generations. Estimating the willingness to pay rests on contingent valuation techniques in which an appropriate and objective sample of the population concerned is asked to indicate the amount they would be willing to pay rather than forego consumption.

Contingent valuation techniques are controversial because they invite respondents to under or overstate their willingness to pay if they believe the outcome of the issue will affect their own interests. For example, a person may be inclined to inflate their willingness to pay, provided they are also confident that they will not personally bear the burden of that payment. If, on the other hand, they are convinced that they will bear the burden, they may be inclined to understate their willingness to pay. Because such funds are normally raised by taxation, a low taxpayer may have a vested interest in a high willingness to pay, knowing that they will bear a disproportionately small part of the burden. There have been cases in Australia and elsewhere

where a small proportion of the sample population accounted for a disproportionately large proportion of the willingness to pay, and these clearly raise doubts about the validity of the evaluation.[45] Much depends on how the questions are stated and whether the respondent knows and understands the other consequences. At this stage, the verdict on this technique is qualified,[46] but it seems likely that it will be pursued in a more rigorously defined manner because there appear to be no alternatives.

The contribution of conservation to the environment is a classic instance of a public consumption good because everyone consumes the atmosphere without generally affecting the capacity of others to do so.[47] Because of this, no market mechanisms exist for charging for the benefits that might flow from increased afforestation or other measures to ameliorate the problem. Investment funds would therefore have to be raised principally from taxes. The scope and orientation of such a program far outstrip commercial interest in plantations which, in any event, have a much shorter time horizon in production than would be desirable for forest established for this purpose. However, the effects of global warming or its amelioration are or will not be uniform for all regions or countries. There will be winners and losers. Hence there are substantial difficulties in evaluating the social net benefits associated with different strategies for amelioration, including those of reducing deforestation and increasing afforestation. Until the predictions are refined to a more precise regional basis, it is difficult to factor this aspect into the issue of conserving native forests other than to note that it is a potentially valid benefit of quite unknown magnitude.

In all cases to date, benefits have been treated without regard to who receives them. As with other forest uses, this often misses the point because considerations of equity may be just as or more important than those of efficiency. If so, it becomes essential to identify the major groups receiving the benefits and bearing the costs. As we shall see in chapter 9, the issue of equity between generations is especially vexing.

The costs associated with managing areas for conservation are substantial and they are critical to the purpose. Contrary to widely held views, areas demarcated for conservation will not simply look after themselves.[48] They need to be protected against illegal clearing, theft, over-use of subsistence edibles or extractives, fire, uncharacteristic exposure to extremes of climate at the edges, imbalances in populations of native animals, feral animals, and invasion by exotic plants, insects and diseases. They need to be surveyed and assessed periodically to monitor present composition and change. At the very least, research is needed to establish the interactions between and the dynamics of plant and animal populations in order to monitor species and population changes. In Australia, comparisons suggest that these costs per unit area are at least half those associated with management for wood production,[49] highlighting the fact that conservation is not a costless forest use.

Another source of costs is the negative or dis-benefits associated with conservation. Not all the contributions of conservation reserves are benign, however much one might wish it was otherwise. Reserves can contribute to illness, injury or death through pollen pollution; by providing a habitat for disease-bearing insects; or through people or animals eating poisonous fungi and fruits or suffering abrasions and cuts from thorns and leaves. They provide a vehicle for mass conflagrations, and injury or death from falling limbs and trees or from attack by animals.[50] Comparisons of net benefits of forest used for conservation with those for other uses need to take account of all benefits and costs.

3 recreation

RECREATION AS A FOREST USE

Recreation in forests takes a variety of forms: some active, such as hiking, orienteering, mountain climbing or canoeing; and some inactive, such as pleasure-driving, picnicking and nature study. Some forms of recreation are concentrated at a particular location or feature and involve large numbers of people; some are dispersed widely and involve very few people at any one location; and some involve several different forms of recreation within the one visit.[1]

Landscape considerations

The role of the landscape and forest environment varies widely in different forms of recreation. Looking from the top of a mountain, or from a plane, the viewer can discern quite small disturbances to the forest, only several hectares in extent, when these are in the foreground, less so in middle ground and still less in the background. From a less elevated position, the ability to discern such disturbances is much reduced and may hinge critically on the angle of lighting or the contrast in line, form, colour or texture involved. The ability to make distinctions also depends on the period of viewing, which may be long for a walker but quite short for a passenger in a car travelling along a road. Similarly with features of scenic beauty, the ability to discern and enjoy may also hinge on panoramic scope, enframement, linear focal cues such as roads or rivers, or features that exhibit contrast in form, colour or texture in addition to these other subtleties. Finally, a viewer's perception of scenic beauty may be influenced by cultural, psychological and socio-economic factors. To cite some examples: migrants from Europe have a different attitude to plantations of exotic pine in Australia from people born in Australia. Tired people may be less receptive to scenic beauty than people who are fresh. Better educated people have a propensity to value scenic beauty more highly than less-educated. People whose recreational activities depend on the form or character of the forest have a more detailed interest in it than do the occupants of cars engaged in pleasure driving; this partly reflects the period of viewing but also reflects the period for reflection.[2] These differences illustrate that forest recreation is a complex and diverse use of publicly-owned native forest, and that generalisations about it are difficult to make or are very weak.

CASE STUDY 3.1 Landscape management of a scenic corridor, Eldorado National Forest.[3]

In managing landscape quality, consideration needs to be given to where and how that landscape will be seen. R. Burton Litton Jr was a pioneer in this field and his work on Eldorado National Forest exemplifies the approach.

Litton and his colleagues carried out a landscape inventory by travelling US Highway 50 between Bridal Veil Falls and Meyers Summit by car and recording critical characteristics directly onto maps or overlays. The 64 km (40 miles) scenic corridor can be traversed in about one hour by a car travelling at typical touring/pleasure-driving speed. In the course of this inventory, foreground, middle ground and background boundaries were noted. Observer position—inferior (below), normal (level), or superior (above)—relative to the landscape was also noted together with descriptions of the landscape types—panorama, feature, enclosed, focal, undergrowth, detail or ephemeral (e.g. seasonal)—and forest types. Figure 3.1 illustrates a portion of the resulting zones, identified serially by number.

For each zone, consideration was given to the possible activities and structures that could influence landscape quality. Prescriptions were developed to maintain or enhance quality by deliberate subsequent management. The descriptions and prescriptions attached to these zones are shown in Table 3.1.

In addition to the inventory summarised in Table 3.1, special attention was also given to the sequence of viewing and its impact and enhancement.

Table 3.1 View categories and prescriptions, Eldorado National Forest

Zone	View category	Management prescription
2.	**Foreground—direct:** Observer inferior, moderate to steep, forested, undisturbed.	Timber—light selection, avoid vertical skid tracks. Roads—retain screen lower side, low standard roads only. Uses—screen uses and key power lines.
3.	**Foreground and middleground:** Observer inferior, moderate to steep, disturbed, fore scars and plantations.	Timber—light selection, fall all dead trees. Roads—promote revegetation and healing of road scars, no new roads. Uses—screen uses, place power lines below ridgetops and no closer than 1.5 km from highway.
4.	**Foreground—direct and peripheral:** Steep to precipitous, forested.	Critical view area No disturbance or use permitted. Remove from timber category.
5.	**Foreground—peripheral:** Steep to forested and undisturbed.	Timber—normal selection, retain hardwoods. Roads—screen from lower side.Uses—screen uses
6.	**Middleground—direct and peripheral:** Moderate to steep, undisturbed, forested, occasional brush fields.	Timber—normal selection to shelter wood on moderate slopes, retain hardwoods. Roads—screen on lower side, avoid large cuts and fills, no road across brush fields. Uses—screen uses, power lines below ridges, minimum clearing no closer than 0.5 km from highway.
7.	**Special feature—foreground:** Rock spire, steep, undisturbed, forested.	Manage as scenic area, allow no visible disturbance to or around vegetation, trails and rock climbing permitted.
8.	**Foreground—direct and peripheral:** Observer superior, steep, critical situation.	Same as 4. Any disturbance will detract.
9.	**Middleground—enclosed landscape:** view with features.	Allow no visible disturbance, roads, power lines or unscreened uses. Timber—salvage harvesting of dying trees only if screened out. No visible skid tracks.

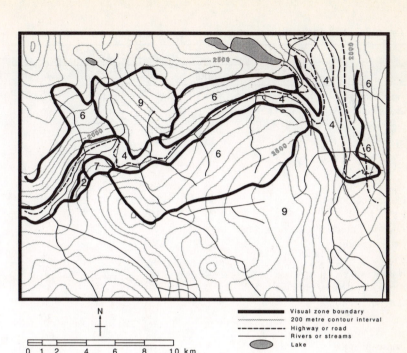

Figure 3.1 Landscape inventory and management, Eldorado National Forest (Litton 1968, Fig. 56).

Activities and factors influencing participation

Data abound on participation rates for various forms of outdoor recreation. The most comprehensive are those based on successive household surveys in the USA and these are summarised in Table 3.2.

The data in Table 3.2 provide some perspective on the relative importance of various recreational activities and the changes over time for one major developed country. However, there are problems in interpreting the detail and trends due to the changing design of the surveys and the inherent variation and potential for substitution in activities,[4] not to mention relating them to forest as distinct from other forms of outdoor recreation. Changing age structures, interests and preferences are probably responsible for the trend in the USA during the 1980s that saw a levelling off and even a decline in visits to some national parks and wilderness areas.[5]

Table 3.2 US participation rates in outdoor recreation activities

	Participation (%)			
	1960	*1965*	*1972*	*1977*
Picknicking	53	57	47	72
Pleasure–driving	52	55	34	69
Sightseeing	42	49	37	62
Swimming	45	48	-	-
Pool	-	-	18	63
Other	-	-	34	46
Walking and jogging	33	48	34	68
Playing outdoor games	30	38	22	56
Fishing	29	30	24	53
Attending outdoor sporting events	24	30	12	61
Other boating	22	24.	15	34
Bicycling	9	16	10	47
Nature walks	14	14	17	50
Attending outdoor concerts	9	11	7	41
Camping	8	10	-	-
Developed	-	-	11	30
Primitive	-	-	5	21
Horse–riding	6	8	5	15
Hiking/backpacking	6	7	5	28
Waterskiing	6	6	5	16
Canoeing	2	3	3	16
Sailing	2	3	3	11
Mountain climbing	1	1	-	-
Off–road driving	-	-	7	26
Visiting zoos etc	-	-	24	73

–, no data available
Data for 1960 to 1977 from Manning (1985, Table 2.1) who collated the respective years of
data from Ferris (1962), Bureau of Outdoor Recreation (1972) and Heritage Conservation
and Recreation Service (1979). Differences in the agencies conducting these surveys may
have contributed to the marked differences between the 1977 and earlier surveys and raises questions as to their comparability.

Early studies of outdoor recreation[6] in the USA found five socio-
economic variables to be related to outdoor recreation patterns:

Age: the older one becomes, the fewer and more passive the activ-
    ities pursued.

Income: the higher one's income, the more numerous the activities
    pursued.

Occupation: the higher one's occupational prestige, the more
    numerous and varied the activities pursued.

Residence: urban residents tend to have higher participation rates
    than do rural residents.

Family stage: the presence of young children tends to reduce the number of outings and makes the recreation pattern more home-centred.

More recent studies tend to confirm these very broad findings but show that other factors, such as cultural differences, can also have a major impact. In ancient Greek and Roman societies,[7] and some more recent, the forest was regarded as a place of evil or bad spirits and this undoubtedly coloured the preferred uses of those forests by those societies. In most societies, different social groups tend to undertake different forms of recreation, as witness the commonly observable differences between the activities of groups of young adult males and those of family groups.[8]

From the point of view of managing the forest, which in some ways may be distinct from managing the recreational activity, the forms of recreation are best categorised by the predominant mode of transport and hence the following descriptive categories are used:

- road-based recreation
- track- and trail-based recreation
- water-based recreation
- forest-based recreation
- eco-tourism.

Road-based recreation

In developed countries, road-based recreation constitutes one of the major recreational uses of forests, in terms of the numbers of visitors, and is centred on pleasure-driving (cars and motor bikes) or riding (bicycles) to enjoy the forest scenery, animals and ambience. Roads are expensive to construct and maintain and many of those used for recreation will have been constructed for reasons other than recreation alone.

Patterns of use tend to follow patterns of leisure time: weekend use being high in forests near major urban centres; holiday period use being high at more distant feature locations. Figure 3.2 illustrates the temporal variation in car visits to a major national park in Victoria, Australia. These temporal patterns of high use can lead to over-crowding. Because many of the roads concerned are public access roads, direct control of visitor numbers is not always possible. Over-crowding on roads slows travel, generally reduces enjoyment of scenery, and may increase the risk of accidents.

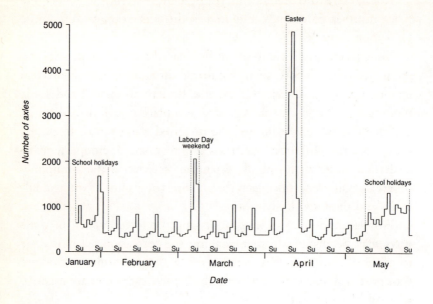

Figure 3.2 Seasonal variations in axle-count measurement, Hall's Gap checkpoint (one-way traffic only), January 1971–May, 1971. After Greig (1977, Fig. 41).

Parking, picnic and toilet facilities normally have to be supplied at appropriate intervals. Signposting, maps of scenic routes and parking at scenic vistas and feature locations also have to be provided for these forms of recreation, not to mention services for maintaining road surfaces, collecting refuse and cleaning toilet facilities. While car-based recreation is not especially demanding with respect to its exclusive use of the forest,[9] other than scenic vistas, focal features and the immediate roadside, the provision of facilities is substantial as is the maintenance required for them.

Track- and trail-based recreation

Four-wheel-drive vehicles and trail (motor) bikes extend the range of vehicular-based use to tracks and trails that are otherwise impassable or unsuitable for conventional automobiles.[10] Mountain (pedal) bikes also use these tracks and trails but are much more benign in their impact on the surface of the track or trail and their generation of noise, and are more limited in diffusion from major access points.

Few if any facilities are provided for these users unless to confine them, as in the case of trail bikes, to a particular location. When four-wheel-drive or trail bike usage becomes intensive through group activities or proximity to urban areas, damage to tracks and trails and associated noise levels may necessitate closing certain areas and restricting these activities to designated tracks and trails.

Water-based recreation

Water has a major appeal as a focal point for many forms of recreation, including passive forms such as picnicking and camping and active forms such as swimming, fishing, rafting, canoeing or boating. Parking facilities are essential wherever the water attraction is accessible by road. These can also be used as a form of rationing by limiting the number of spaces available, or of charging where usage rates would support the costs involved. Toilet and picnicking or camping facilities are needed where water-based recreation is concentrated on lakes, dams, inlets or large rivers. Conflicts can arise between motorised and other forms of aquatic transport because of noise and safety concerns. Motorised forms of aquatic transport may also damage land at the edges of the water unless they are restricted in speed and number.

Forest-based recreation

In developed countries, walking, cross-country skiing and orienteering are some of the most popular forest-based uses. Horse-riding and hunting are further uses but are very much dependent on the terrain and vegetation or animals present and whether or when hunting is permitted.

Walkers, especially the more active and adventuresome, often have a preference for solitude relative to other group and human interference, and for naturalness. These properties are associated with walks taking several days through large tracts of old-growth forest or so-called wilderness. Wilderness recreation has become one of the most vexed issues because it is not only demanding of space, but it places a considerable premium on solitude and naturalness for the walker or group of walkers.[11] Cross-country skiing is somewhat similar but is obviously seasonal in character and reliant on suitable conditions.

Orienteering tends to be restricted to forests that are relatively easy to run in, with little undergrowth. It is an activity with very low impact on the forest. Most horse-riding, on the other hand, is confined to trails and is generally not compatible with other uses such as walking. Horses are also potential spreaders of weeds and, like other hard-hoofed animals, can cause damage to trails unless drainage and surface-hardening are maintained.

These forms of recreation are relatively undemanding in terms of the facilities that have to be provided. Indeed, the absence of facilities may be important to the recreational experience. Crowding from unrelated visitors or groups, on the other hand, generally detracts from the recreational experience and various devices for rationalising use may have to be implemented to control this problem.

Eco-tourism

In many countries, both developed and developing, special attention is being given to feature-based recreation catering for international or national tourists.[12] These are visitors who are prepared to pay high entrance or participation fees for a relatively short but intensive and different recreational experience. The features on which these eco-tourism ventures are based vary widely in character and may be road, water or trail based. Examples include sightseeing by road, elevated walkways in the canopy of rainforest; adventure tours in four-wheel-drive vehicles, raft travel on wild rivers, skiing or mountain treks, and wildlife safari tours. All involve substantial investment in access and facilities to cater for tourists mainly drawn from upper income and better educated classes, although young well-educated but low-income backpackers and older retirees represent significant and different groups.

BENEFITS AND COSTS OF RECREATION

Where use is or can be concentrated on defined access paths such as dedicated roads, parking areas or gateways, recreation is potentially a market good because an entry or use fee could be charged. If the level of use is insufficient to cover the costs of collecting fees, which is often the case with forest recreation, free entry may be given and recreation is thus a public consumption good, unless overcrowding occurs. If it does, the public consumption good status may be changed to an unpriced market good, since the consumption by one person may affect the capacity of others to consume and enjoy. This illustrates the sometimes subtle distinction between (unpriced) market and public consumption goods.

In any event, even where prices are charged they may not represent a good reflection of a market price because of inherent spatial and other imperfections of the market for recreation. Spatial imperfections include those associated with the site-specific character of many recreational activities, the consequent lack of alternative suppliers, and the concentration of consumers in urban residential areas. Other imperfections include those of imperfect information and lack of competition or, to use a less demanding term, contestability. Much of forest recreation in public forests is administratively rather than truly market priced and much of the remainder is unpriced.

For these reasons, prevailing prices for forest recreation may not represent useful measures of the benefits of recreation. Other techniques have therefore been developed for estimating willingness to pay in order to estimate the benefits of recreation.

Travel cost method

The most widely used technique is that of the travel cost method.[13] Based on an objective survey, forest visitors are classified into different geographic zones of origin. The imputed travel cost is then regressed against the relative visitor use (relative to base population) for each zone. On the assumption that visitors would react to an additional dollar spent on recreation in the same way that they reacted to an additional dollar spent on travel, it is then possible to derive a demand curve for recreation from this estimated relationship.

Cross-section analyses

Another technique used to estimate the demand forms of recreation is based on the analysis of cross-section data collected from household surveys.[14] These surveys provide household data for annual participation, average cost, household income and other household variables. Regressions of participation against cost (as a proxy for price) and other household variables provides an estimate of the aggregate demand for that form of recreation. From that function, the willingness to pay can readily be evaluated. However, this is an aggregate value for the form of recreation, not one specific to a site.

Contingent valuation—users

Contingent valuation techniques were outlined in chapter 2. Several comparisons have been made between travel cost and contingent valuation estimates for users of the same site. The results are far from clear. All one can say is that they are in broad agreement as to the order of magnitude but quite unclear about which is to be preferred as both have problems.[15]

Contingent valuation—non-users

What of the non-users? One can raise the argument that non-users may have an existence value for recreation as for conservation. This is virtually inseparable from that for conservation and subject to the same debate. Recent reviews[16] have cast a shadow over most past research using contingent valuation because biased or methodologically inappropriate techniques were used. On the other hand, these reviews have identified the conditions and methods under which valid future studies might be conducted. These conditions are exceedingly demanding of design and administration and will limit application of this technique. However, it is currently the most promising method for obtaining valid estimates of the benefits to non-users.

Costs

The costs associated with providing forest recreation are often incurred jointly with other activities such as conservation or wood production. As noted earlier, this makes it imperative to identify a cost-driver or, failing that, to allocate subjectively between the uses on the basis of the perceived relative contributions to incurring the costs. Some recreational uses of roads, for example, arise as unintentional by-products of harvesting operations. In such a case, it is generally not legitimate to charge the initial capital costs of the road although it may be appropriate to charge some of the maintenance costs because of the additional recreational use of those roads.

CASE STUDY 3.2 Economic benefits and costs of management of Khoa Yai National Park, Thailand.[17]

Khoa Yai National Park is located at the southwestern edge of the Khorat Plateau in northeastern Thailand, about 290 km northeast of Bangkok. It covers 2168 square km and was established in 1962, being Thailand's first national park. Despite its protected status, forest cover has declined from 94 per cent in 1961 to 85 per cent in 1985, due principally to illegal clearing for agriculture. Some poaching of timber, wildlife and other forest products also continues.

The principal benefits from the park are derived from its conservation values relating to biodiversity, including rare wildlife and substantial areas of relatively undisturbed ecosystems that are becoming increasingly rare, from recreation, and from water supply. The park may harbour as much as 10 per cent of Thailand's remaining population of elephant, gaur, tiger and pileated gibbon. Visitor numbers increased from 115 565 in 1977 to 401 661 in 1987; an estimated 95 per cent being Thais. Relaxation, wildlife viewing and the scenery appear to be the major attractions. The park has lodge-type and motel accommodation but about 25 per cent of visitors engage in overnight camping. At present, the park watersheds have a very low erosion rate, 0.7 t/ha/year, thanks to the forest cover.

The benefits from conservation are difficult to value. Some attempt at contingent valuation has been made, yielding an estimate of average willingness to pay to ensure the continued existence of elephants of US$7.00 per visitor, but is subject to the many doubts and perceptions of bias that attach to all earlier applications of this technique.[18]

Estimates of recreation expenditures span a wide range, from US$1.5 to 7.7 million per year, depending on the assumptions made. These include transport, accommodation and food, and mostly accrue to the tourist authority, tour companies and food providers. Entry fees amount to about US$110 000 per year.

Estimates of water values are highly speculative, being based on the possible erosion rates that might be achieved if forest cover was removed. Even so, they do not appear to be of major consequence for one of the two major downstream reservoirs. The other has already suffered major incursions from other sources but might incur substantial costs if uncontrolled clearing or harvesting were allowed on a large scale.

The variable costs of managing the park amount to about US$130 000 per year. The capital works budget varies widely but was about US$70 000 per year in 1986 and 1987. This suggests that the park is running at a substantial financial loss. This loss would increase greatly if the substantial additions to staff and accommodation proposed in the 1987 Management Plan were to be implemented. Several ways of redressing this financial loss have been advanced, notably to institute different entry fees for overseas tourists[19] and for the park service to take over the tourist authority lodges to provide a larger commercial base.

There are also two major sources of opportunity costs involved. The first concerns the timber resource, amounting to some 50 million m^3 of commercial species. Discounting this by half for necessary protection reserves and applying 100-year rotation, this might yield an annual net revenue of at least US$1 million per year.[20] In addition, villagers in surrounding areas incur an opportunity cost of the loss of access to other forest produce. After deducting that portion gained from illegal poaching of elephants, this was estimated to be about US$1 million per year.

These statistics are extremely imprecise but they highlight the difficulties that confront large-scale reservation of forest areas in developed and, even more acutely, in developing countries. The issue is not whether or not there should be a park. Very few would query that in this particular case. The financial burden posed directly through management costs in excess of revenues (and much more so indirectly through opportunity costs), together with the difficulty of evaluating the non-monetary benefits, make responsible economic management of forest land extremely difficult even when some of the complexities of joint production of wood and non-wood uses are absent.

4 water

WATER AS A FOREST USE

Water has a number of uses besides those of a consumptive character, including being a medium for transportation, a habitat for aquatic wildlife, and a recreational venue. Transportation is still a significant use in a number of countries, but is relatively undemanding in character and need not be considered separately. Apart from swamps and lagoons, and certain physical requirements such as fish ladders on dams and shade from fallen logs on streams, the habitat requirements for aquatic wildlife are generally similar to those for consumption. Recreation has already been considered in the preceding chapter and, although the role of water as a recreational venue is not as demanding as some of the consumptive uses of water, the principles involved are similar. Consumptive uses of water play a far more critical role in determining the quality and quantity issues associated with water production from forests and thus form the focal point of this chapter.

Flood control is the other dimension to water use and management in forested catchments. Rapid snow-melt or sustained and widespread high levels of precipitation can generate floods that damage downstream land, buildings, bridges, roads and other property and resources and can endanger life. Flood control measures to increase channel capacity or reduce the rate of flow mostly involve investment in downstream works, such as the construction of channels or artificial dams. These are largely outside the domain of public forest management and will not be pursued further[1] here. The rate of flow can be modified, to some degree and under some circumstances, by management of the forest itself. The measures and effects are common to those for consumptive use and no distinction need be made here.

Water is collected for consumption in above-ground artificial reservoirs or natural lakes, or in artesian or sub-artesian reservoirs, or by pumping or diversion from free-running rivers and streams; the first and second categories being the most common forms for urban communities today. Forested catchments play a dominant role in maintaining both the quantity and quality of the water inputs into many of these systems because they occupy higher rainfall areas and are relatively free from contaminating influences.

The three major properties[2] of forested watersheds mostly responsible for the quantity and quality of water inputs into consumptive systems are:
- interception,
- evapo-transpiration, and
- infiltration.

The conversion of forest to agriculture, fire, and human and animal diseases represent other significant influences.

Interception

As noted earlier, forests may be significant in increasing interception of atmospheric water in clouds, fogs or mists through the roughness of the canopy and boles and their additional height. The montane cloud forests are the best known examples, but there are many other localised examples such as the redwood (*Sequoia sempervirens*) forests of the coastal fog belt of California and Oregon.[3] In chapter 2, reference was made to the role of the Amazon Basin rainforests in increasing rainfall over the inland part of the Basin; there may be other examples of lesser extent, including the Congo Basin forests.[4]

In drier climates, the converse effect may be apparent. Forests established or re-established on grasslands may increase interception losses through evaporation from leaves, branches and boles before rain can reach the ground.

Interception is generally not a property that can be deliberately or directly manipulated in the interests of increasing the quantity of water available on forested catchments, although the rate of melting

snow and hence the duration of peak flows and likelihood of flooding can be influenced by the extent of forest cover.

Evapo-transpiration

Evapo-transpiration, canopy cover and shade and shelter are generally complementary in character—the denser the canopy, the more the
shade and shelter and the more the evapo-transpiration, since all are
dependent on leaf surface area. The quantity of water inputs into
storage systems can be manipulated by varying the density or age of
trees to take advantage of evapo-transpiration and shade effects.

Short-term effects relate to the capacity of the forest canopy and
tree roots to reduce the surface flow and thus the rate of release of
water to watercourses. Catchments with permeable subsoil associated
with deep-rooting forests may delay the peaking and rate of release
of water from extreme events. This is a significant property in regions
with a dry summer because it spreads the input into storage over a
much longer period and thus may reduce the downstream flow in
large catchments during spring, ameliorating possible flooding.
However, catchments are complex systems. If a storm is short and the
past weather dry, there may be considerable buffering capacity. Only
a small part of the catchment may be effective in generating surface
run-off and subsoil through flow. As the extent of a storm increases,
more and more of the catchment generates run-off and through flow,
until there is no buffering capacity and the total rate of release
approaches the rate of precipitation.[5]

Long-term effects relate to the changes in evapo-transpiration
with age or with canopy cover. Put briefly, once beyond the seedling
stage, young trees transpire more than old trees and dense forests
transpire more than open forests,[6] leaving less water to proceed to
storage systems. These effects can be substantial and will be considered further later in this chapter.

These long-term effects also need to be related to the patterns of
rainfall, which vary greatly in intensity and frequency. Satterlund and
Adams[7] cite a vivid example of the resulting variability in streamflow.
In March 1913, the Sacaranga River in New York reached its highest
recorded rate of flow of 906 m^3/s, only to decline by September of
the same year to 0.15 m^3/s. The study of forest hydrology is therefore
inevitably one that deals with probabilities rather than absolutes.

Infiltration

Infiltration refers to the water transported as through flow within the underlying soil or rock, as distinct from surface run-off. Infiltration is principally an indirect determinant of water quality, because the lower the infiltration rate the higher the surface run-off. Higher surface run-off leads to more soil erosion. Dense forest has a dense litter layer and root mat which tend to resist surface erosion. The root mat tends to aid infiltration through improved soil structure and by transporting water through decayed root channels.[8]

However, erosion and water pollution from point sources are generally far more significant than from non-point or general surface flows. Thus roads, trails and tracks of any sort tend to be focuses for point pollution in forests because they concentrate the lines of drainage. Other more localised sources of point pollution are the edges of streams, lakes or dams or other sites used intensively by humans or animals.

The difficulty in assessing the effects of these point or non-point sources of pollution is that erosion is also a natural process; it is always present, independent of humans or animals, but tends to be episodic and very variable. Human or animal-induced pollution is also influenced greatly by the magnitude of the rainfall event. Thus one is comparing inherently probabilistic or stochastic processes for which little information is generally available, especially regarding the background level of natural erosion events.

Conversion to agriculture

The consequences for water quantity and quality of converting public forests to agriculture are not a major issue in developed countries because those patterns of land use are relatively stable. However, it remains one of the major problems confronting developing countries. The fringes of public forests are constantly being eroded by clearing or illegal burning, often on increasingly steep slopes, as peasant farmers pursue fuelwood supplies or extend their cropping areas.[9] In some countries, these activities have not been confined to peasant farmers; in Brazil, for instance, earlier subsidies for clearing rainforest for cattle grazing were responsible for extensive conversion.[10]

Any access road in public forest offers the opportunity for illegal conversion along itself, whether for shifting or more permanent agriculture. Forest and park services are frequently unable to control these activities, either because of a lack of staff and resources or because the political pressures for land make control measures infeasible.

If clearing is combined with fire on steep slopes, the effects on water quality are often severe. The effects on quantity are less marked immediately, but may become significant as clearing progresses further and further over a period of years.

Fire

Fire, whether caused naturally or by humans, can have a substantial impact on water quality and quantity if it is severe, and may kill the canopy or a substantial part of it. The effects of a loss of canopy and evapo-transpiration lead to a greater surface run-off, which may be exacerbated by the hydrophobic properties of burnt mineral soil. Thus discharge rates are faster and rise more quickly after extreme or even moderate rainfall events than is the case for adjacent similar unburnt catchments. Levels of ash-derived soluble elements, such as sodium and potassium, are also increased in run-off.

CASE STUDY 4.1 Discharge rates of water and nutrients following wildfire near Powelltown, Australia.[11]

On 16 February 1983, a high intensity wildfire burnt about 40 000 ha of predominantly mountain ash (*Eucalyptus regnans*) forest, including a 35 ha sub-catchment near Powelltown, in southeastern Australia. On 22 February 1983, an intense thunderstorm in the locality caused a mud torrent across the road traversing the base of the relatively narrow catchment. Subsequent surveys indicated that a high proportion of the slopes had been sheet-eroded, and the single gully drained by an ephemeral stream had been extensively scoured by high flows.

The soils of the catchment are porous, with a high infiltration capacity under vegetated (unburnt) conditions, and would normally prevent overland flow except perhaps during long duration storms of high intensity. Prior to the fire, the vegetation on most of the catchment probably comprised an overstorey of mixed eucalypts (*E. cypellocarpa, E. sieberi, E. baxteri and E. obliqua*) and associated understorey species, with an estimated 100 t/ha of fuel in the litter and understorey. Fire intensity varied from a light back-burn near the road to a crown fire in 1–2 ha patches on the steeper slopes. Visual assessment indicated around 25 per cent of crowns in the catchment were consumed, 70 per cent were scorched and killed, and 5 per cent were unburnt. All understorey and litter fuels were consumed on 33 ha of the 35 ha catchment. While the ash surfaces were highly water absorbent, the underlying soil to a depth of 10 cm was hydrophobic, a property believed to be drought-induced but enhanced by the intense heat of the fire.

Deposits of charcoal in the drainage line where it crossed an under-ground pipeline indicated a peak flow of between 5000 and 10 000 L/s, corresponding to rainfall intensity of around 50 mm/h for only 10 minutes, which accords well with typical rainfall events at the time of year. Flows of this level are sufficient to account for the mud torrents. Of the 33 ha of catchment burnt, 38 per cent was not eroded, 58 per cent had been eroded of ash and loose soil, and depositions covered about 4 per cent of the area. Most slopes of greater than 20° were sheet eroded, but lesser slopes only eroded where overland flow from higher slopes occurred.

Total losses of nitrogen and phosphorous were conservatively estimated to be 225 and 11 t/ha respectively. While small relative to the total values in the ecosystem, these represent about one-third of the above-ground bio-mass. Thus wildfires of this nature probably exacerbate the fertility gradients that are already evident between the ridge and gully sites in this area.

Human and animal diseases

The final major determinant of water quality stems from the intro-duction of water-borne disease by humans or animals. The best known case is that of Yellowstone National Park in the USA, where campers and walkers are said to be responsible for *Giardia* becoming rampant throughout the waters of the park.[12] Most water supplies for domestic consumption in developed countries are now chlorinated to eliminate diseases. Those that are not chlorinated are normally closed to humans and domestic animals.

BENEFITS AND COSTS OF WATER

Water used by humans, whether consumed directly or used for electricity generation, manufacturing, irrigation or stock, is a market good. Public and private water authorities charge for water use. Because of the natural monopoly power inherent in the spatial management of water supply, the prices charged often bear little relation to market transactions in that domestic water is frequently charged according to property rates rather than actual consumption, and discriminatory pricing is often applied to commercial versus domestic consumers of water or electricity. Some of these imperfections in pricing practices are being redressed and reasonable estimates of the price of water (or electricity) delivered to consumers are often available. But these are not the prices appropriate to the valuation of water at the storage facility because they incorporate the costs of reticulation, and, where appropriate, of the water storage and treatment facilities. While it is possible to develop cost-benefit analyses of the decision to invest in these facilities, using these prices and taking account of all the costs involved, this sheds little light on the value of water at the storage facility.

The demand for water at the storage facility is a derived demand. It is therefore theoretically possible to estimate the price of water at the storage facility by subtracting the long-run marginal cost per unit of reticulation and associated facilities from the imputed market price to the final consumer. In this respect, the process is identical to that used in deriving a price of timber standing on the stump from that of timber sold to the builder. However, these calculations are even more precarious because the investment costs in facilities are very large and the imputed market price of water to the final consumer is most uncertain. Small variations in investment costs or imputed market price can shift the derived demand price from a high positive value to a high negative value. Hence this approach is seldom used for valuing water although, as we shall see, the principles are applicable to deriving scarcity prices for water from a catchment.

A simple model of pricing water as a scarce resource at the catchment level can easily be developed. The argument rests quite simply on a theorem from the theory of welfare economics, that a public monopoly such as a water supply authority should price at long run marginal cost in order to maximise net social benefit .[13]

Let us assume that we are dealing with a domestic water supply authority supplying a steadily growing metropolis in the long run. For simplicity, let us assume that consumption has already reached the capacity of the first dam to supply. Long run average cost is constant over the possible range of supply from it and thus equals the long-run marginal cost. Looking ahead over two incremental jumps in time, population and water consumption, each successive dam site involves progressively higher and higher construction costs if, as is reasonable to assume, the authority has investigated and ranked sites in order of cost per unit of water supplied.

At the time the capacity of the first is just exceeded by consumption, the authority must price at the long-run average cost of the second dam.[14] This will be higher per unit of water than the earlier figure because the second site is less favourable. At this time, the value of the water at the second catchment is zero, but that at the first will be earning a rent equivalent to the difference in average (or marginal) cost of water from the two dams. The argument can be extended to the third dam such that the first two will be earning rents equal to the differences between their respective average (or marginal) costs and that of the third.

As with other goods with scarcity values or rents, there is a limit to the magnitude of the value or rent at any point in time that is set by the alternative means[15] of satisfying the demand for water. They include augmentation through diversions, raising dam walls, desalinisation, treatment or recycling, and more efficient technologies of use on the supply side.

Figure 4.1 illustrates the elements involved in determining the price of water and shows a hypothetical aggregate demand curve DD', and supply curves $S_cS'_c$ and $S_cS'_b$. The supply curve $S_cS'_c$ shows the incremental steps in long run average (and marginal) cost associated with each new catchment, and the role of backstop supply in $S_cS'_b$. The value attributable to water at the catchment, assuming no timber-harvesting, is the producer's surplus or rent. On a unit value basis, this translates to the difference between the price charged to the consumer and the long run average (or marginal) cost per unit of water for the catchment concerned; P_1 and P_2 respectively for the first and second catchments on figure 4.1.

In practice, comparisons of actual long run average costs of two dams constructed at different points in time may be distorted by inflation on one hand and technological changes in construction on the other. However, it is relatively simple to obtain an estimate of the current cost of construction, using current prices and technology for each future dam or augmentation, and so determine the average costs concerned.

This simple analysis of scarcity prices of water demonstrates that different prices for water will be appropriate to different catchments or augmentations of catchments. If a catchment supplying a dam carries old-growth forest and is the first to be constructed, such that the scarcity value of water from it is high, there may be a clear case for retaining that forest cover, at least until fire or other catastrophe occurs, in the interests of increasing water production. The outcome will depend on the actual scarcity price of the water and the resulting

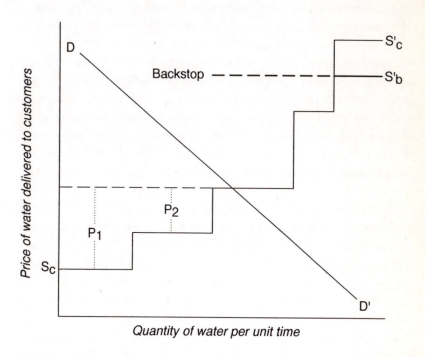

Figure 4.1 Hypothetical supply and demand curves for water delivered to customers

net benefits from water and the net benefits from timber production: an analysis that can be pursued using simple analytical models.[16]

CASE STUDY 4.2 Joint production of water and wood from the Thomson Catchment.[17]

Melbourne is the capital city of the State of Victoria in Australia and currently has a population of about 2.5 million people. Its water supply comes from forested catchments to the east of the city that are characterised by a Mediterranean climate with a dry summer and occasional droughts following low rainfall in winter.

Research on water yields in the Thomson Catchment indicates that significant losses of water production are likely to follow clearfelling of forest due to the consequent higher transpiration losses from the young regrowth forests, possibly rendering joint production uneconomic. These losses diminish as the forest grows older, especially beyond sixty years of age.[18] The reduction in water production has prompted considerable debate over the appropriateness of continuing to allow the Thomson Catchment to be used for timber production as well as water production.

Table 4.1 shows past and future augmentations to the Thomson Catchment and other catchments supplying Melbourne, and the capital and other costs associated with those augmentations. Amortising these capital costs over 100 years and adding operating costs leads to estimates of the long run marginal (and average) cost for each augmentation. These in turn enable estimates of the scarcity price of water *from the Thomson Catchment* to be derived at various dates in the future, based on the principles outlined in figure 4.1

The long run marginal cost curve traced out by the data in Table 4.1 does not follow the classic textbook pattern shown in figure 4.1. While it is possible that an aberration has occurred in the planning process, whereby the successive development of augmentations does not follow the rational pattern of progressively increasing costs, the more likely explanation is that some, if not all, of the decreases reflect contingent economies of scale. This is patently the case for the sequence of Macalister Catchment augmentations, where an initial higher capital cost is required in order to provide an appropriately engineered foundation for the later stages. However, the extraordinary increase associated with the construction of the Lower Yarra, Watson's Creek augmentation remains an anomaly because there is no obvious reason why it should

Table 4.1 Scarcity prices of water from the Thomson Catchment

Augmentation (new storage)	Approximate timing (year)	Capital cost ($M)	Storage supply capacity (GL/yr)	Annual interest & depreciation ($/ML)	Long run marginal cost ($/ML)	Scarcity price of water ex Thomson ($/ML)
Thomson dam	1992	900	220	206	254	0
Big River diversion	2010	129	80	81	129	0
Black River diversion	2019	179	68	133	181	0
Black River storage	2028	146	28	263	311	57
Lower Yarra Watson's Creek storage	2032	468	52	453	542	289
Macalister storage stage 1	2039	734	97	381	429	175
Macalister storage stage 2	2057	495	80	312	360	106
Macalister storage stage 3	2075	945	170	280	328	74

Source: Ferguson (1995).

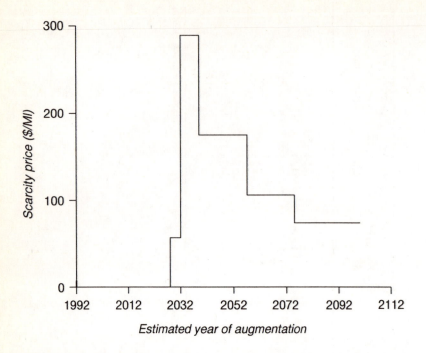

Figure 4.2 Estimated scarcity price of water from the Thomson Catchment in future years (from Table 4.1).

be constructed prior to the Macalister augmentations. Theory and reality often differ: that does not negate the principles involved.

Using the data in Table 4.1, the supply price for water *from the Thomson Dam* over time can readily be derived and is shown in figure 4.2.

Up to the time of construction and use of the Black River storage, no calculations of the economics of joint production of water and wood are needed because the price of water from the Thomson Catchment remains at zero: the differences in the successive long run marginal costs for the Big River storage and Black River diversions being *negative*. and thus there being no scarcity value. Joint timber and water production is the optimal solution, at least until 2028. The later economics can be evaluated using a simple analytical model.[19] Timber production remains economic until the Lower Yarra Catchment augmentation is operational. From 2032 to 2039, or whatever the date of use of the Lower Yarra Catchment augmentations, the price of water from the Thomson Catchment is high and the opportunity cost of water losses due to regrowth transpiration is therefore also high. Timber production is not optimal and should cease unless ways of reducing the water losses by strip felling or thinning can be found. Beyond 2039, the price of water declines with each successive augmentation and joint timber and water production would once again become optimal by the last augmentation shown.

The same principle of pricing pertains to artesian and other forms of storage or supply. As consumption reaches the capacity of the existing system, new and (generally) more expensive sources must be tapped and the original water catchment then earns a scarcity rent. However, water from a new and more expensive storage facility is at the extensive margin of production and thus earns no scarcity value or rent.

The costs of managing water catchments for water production alone are those of protection against fire, feral animals and pests (which may pollute water or damage forest cover) and such other deliberate activities to increase water yields or quality as may be desirable.

As with other goods, water is not always benign. Natural floods can cause damage and impose costs on individual landowners, occupiers and businesses. These impose external costs that have to be taken into consideration in appropriate circumstances. More importantly, they set a base level against which human-induced or potentially induced floods must be gauged. As with other goods and services, the difficulty is to establish the background level of risk associated with entirely natural occurrences in the past. Data for this purpose are scanty and at best date back some fifty years or so. Viewed in the longer perspective, this length of time is a quite inadequate sample that only reflects a very small part of the current interglacial period and its climate.

5 wood

WOOD PRODUCTION AS A FOREST USE

Wood is used in a variety of forms and for a variety of purposes. In many developing countries, fuelwood and simple construction materials such as poles dominate the patterns of consumption. In those developing countries with substantial forest resources, logs and partly processed wood products such as rough-sawn timber and veneer may be exported in large volumes. In developed countries, the emphasis is much more on manufactured products used for building or for paper or packaging.

Official FAO statistics on the ownership of native forests are at best patchy, making it difficult to relate wood production and consumption data specifically to the focus of this book—the issues relating to publicly-owned native forests. Of necessity, ownership matters are thus based on first hand experience, anecdotal evidence, and a smattering of data. Publicly-owned native forests used to supply fuelwood and poles in the developing countries with relatively poor forest resources suffer different pressures from the well-endowed countries supplying industrial wood. Hence it is desirable to deal separately with the present and future patterns of production and consumption for these different groups of products.

Fuelwood and poles

Statistics on fuelwood and charcoal consumption are summarised in Table 5.1.[1] These include the roundwood equivalent volumes of charcoal but exclude poles.

Table 5.1 Production and trade statistics for fuelwood and charcoal, 1992

	Fuelwood and charcoal volume (million m³/yr) †			
	Production	Imports	Exports	Consumption
Developed countries				
Europe	53	3	1	55
Former USSR	81	-	-	81
North America	100	-	-	100
Australasia	3	-	-	3
Japan and others	8	1	1	8
Sub-total	245	4	2	247
Developing countries				
Africa	449	-	-	449
Asia	879	1	1	880
Latin America	296	-	-	296
Other	5	-	1	5
Sub-total	1629	1	2	1873
World total	1873	4	4	1873

† Volumes stated are roundwood equivalents
Data source: Food and Agriculture Organisation (1994).

Table 5.1 illustrates the striking difference between the developed and developing countries in their dependence on fuelwood. The addition of poles to these statistics would shift the balance slightly away from the developing countries. However, in developing countries poles are used predominantly for simple building structures, whereas those in developed countries are used predominantly for transmission lines, so the uses and prices are very different.

As Table 5.1 shows, only small volumes of fuelwood and charcoal are traded internationally. Trade in building poles is negligible, but there is a small but significant trade in transmission poles, principally from developed to developing countries.

Fuelwood is a bulky heavy commodity with relatively low unit value, making shipping and international trade unattractive. Nevertheless, substantial volumes are carried by ground transport within some of the countries concerned, notably in Africa, as witness bundles of fuelwood strapped on unlikely places on trucks by drivers engaged in wood supply as a private sideline. Here, supply zones for fuelwood may extend many hundreds, if not thousands, of kilometres from the centres of population and domestic wood use. There are also many remote areas in Africa and elsewhere where transport is almost entirely by human ('head-loading' in India) or animal, imposing more stringent limits on the extent of the supply zones. A general limit of about 6 km prevails for head-loads but there are fuel-scarce provinces in the Indian sub-continent,[2] China and Africa where much greater distances are travelled, generally by women.

As figure 5.1 shows,[3] the annual production of fuelwood and poles is expected to continue to increase fairly rapidly at 1.6 per cent per year in developing countries to the year 2010, the rate of increase being directly linked to population increases in the developing countries and the lack of alternative cheap fuels. The dependence of many developing countries on fuelwood and charcoal needs to be stressed. In Africa, it represents 36 per cent of the total energy supply, 20 per cent in South America, 46 per cent in Indonesia and 25 per cent in India.

Most fuelwood and pole production is from publicly-owned native forest, with the exception of extensive private fuelwood plantations established in Brazil. In general, however, the prices received for fuelwood and poles have not been conducive to private

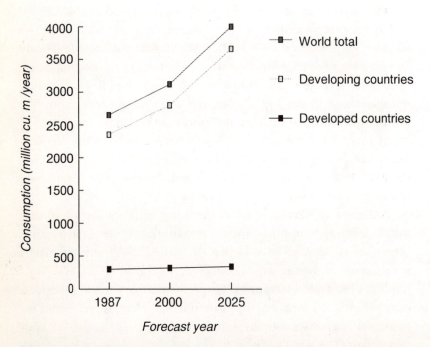

Figure 5.1 Forecasts of consumption of fuelwood and poles (Sharma et al. 1992, Exh. 2–8).

investment in plantations or other forms of fuelwood production. In those countries where fuelwood is scarce, control of harvesting on publicly-owned native forests becomes the dominant issue. In earlier times, control measures were based on policing the regulations with the aid of financial and other penalties. This seldom proved sufficient or effective where demand from surrounding villages was high. Increasing emphasis is now being given to community education to enlist the involvement and support of local populations in regulating this use of publicly-owned native forest, often under the rubric of 'social forestry'. Where fuelwood and building timber is scarce, wood production is generally a primitive process involving short rotation lengths and scant consideration of any other forest values or uses. Fortunately, many of the species coppice and thus can potentially sustain a continuing supply of wood.

Industrial wood

Wood products are numerous but diverse. Some are primary products from processing logs; others wholly or partly utilise wood residues from other forms of processing or residues that would otherwise be left following harvesting operations. Figure 5.2 illustrates the flows and linkages that apply in many markets for wood in developed countries.

Statistics for production and trade in industrial wood[4] are summarised in table 5.2. Industrial wood includes all wood harvested and traded in the round, plus the roundwood equivalent volume of pulpwood traded as woodchips, but excludes fuelwood and charcoal.

As Table 5.2 shows, the developed countries dominate the pattern of production and trade in wood as a raw material. Restrictions have recently been imposed by many developing countries on log exports and have reduced the total volume and value of exports considerably. Asian imports have been expanding rapidly, however.

Figure 5.3 shows the forecasts[5] for future production to the year 2025. These forecasts involve very substantial increases for the developing countries as well as moderate increases for developed countries.

Table 5.3 summarises the production and trade statistics[6] for these final products, based on the combined volume of sawnwood and veneers to represent solid wood products and the mass of wood pulp to represent paper products.

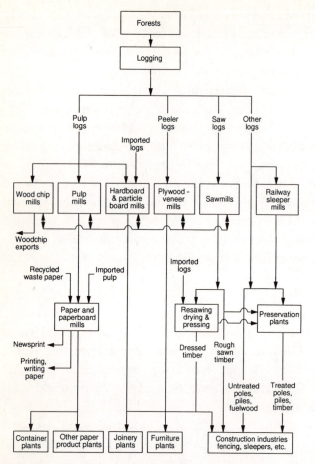

Figure 5.2 The flow of wood between the processing industries (Ferguson 1985, Fig. 3.1).

Table 5.3 shows the dominance of the developed countries in production and trade of wood products, often both as importers and exporters, as might be expected with the wide array of finished products that flow from these intermediate goods. However, the Asian region is a significant importer and exporter of sawnwood and is likely to become a major net importer of wood pulp and paper, notwithstanding the major increases in pulp production that are in train or have been completed in Indonesia in recent years.

Forecasts[7] for future production levels of wood products suggest modest increases for sawnwood in developed countries and somewhat higher rates of growth for wood pulp production, very rapid

Table 5.2 Production and trade statistics for industrial wood, 1992

	Volume per annum (million m³/yr) †			
	Production	Imports	Exports	Consumption
Developed countries				
Europe	278	41	2	317
Former USSR	256	-	11	245
North America	582	4	31	555
Australasia	32	-	11	21
Japan and others	40	47	29	58
Sub-total	1188	92	84	1196
Developing countries				
Africa	44	-	-	44
Asia	240	21	1	260
Latin America	128	-	-	129
Other	3	-	1	2
Sub-total	415	21	2	434
World total	1603	113	86	1630

1† Volumes stated are roundwood equivalents
Data source: Food and Agriculture Organisation (1994).

Table 5.3 Production and trade statistics for wood products, 1992

	Sawnwood and veneers (10³ m³/yr)			Wood pulp (10³ t/yr)		
	Production	Imports	Exports	Production	Imports	Exports
Developed countries						
Europe	67	35	28	35	14	8
Former USSR	52	1	3	8	-	-
North America	137	35	56	82	5	15
Australasia	4	1	1	2	-	1
Japan and others	26	11	-	13	3	-
Sub-total	286	83	89	140	22	24
Developing countries						
Africa	1	2	1	1	-	-
Asia	24	9	9	5	4	-
Latin America	15	2	2	8	1	3
Other	-	-	-	-	-	-
Sub-total	40	13	12	14	5	3
World total	326	96	101	154	27	27

Data source: Food and Agriculture Organisation (1994).

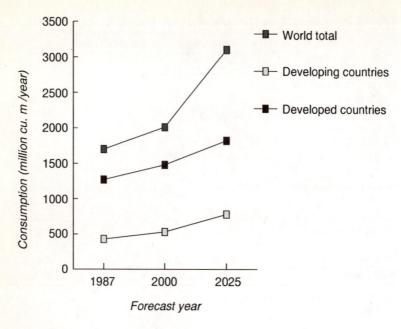

Figure 5.3 Forecasts of consumption of industrial wood (Sharma et al. 1992, Exh. 2–8).

increases being expected in developing countries. Predictions of the market outcomes in terms of demand and supply shifts and trade patterns are fraught with difficulties. However, the general consensus for the period to 2010 appears to be one of increasing pressures on supplies of sawnwood and wood pulp, punctuated no doubt by the cyclical swings that characterise both markets.

Claims of forthcoming timber famines need to be discounted. First, the forecasts themselves are imprecise and many factors may change the outcomes. Second, industrial wood is widely-grown and traded and producers and consumers respond to the price signals of the market. If supply declines or demand increases in the long run, prices will rise, and this will normally encourage more investment in growing wood. True, the long period of production involved causes time lags in the process and may allow substitutes to enter the market, but these are the normal complications of markets for a good with a long production period. This is a very different situation from that of fuelwood and poles, where poverty, population increases and the absence of inexpensive substitutes may give rise to local wood supply shortages despite the shorter production period (eight to twenty years) that is generally feasible for these products.

In developing countries where export of wood is or was important, the problems are of a different character. Here, the prospect of revenues from exports and of payment of stumpage or license fees to the state as owner of the public forests, makes the wood products sector an important part of the cash economy of the state, at least for a time. Governments in developing countries hard-pressed for foreign exchange, find it difficult not to look at a major native forest resource simply as a source of capital and foreign exchange waiting to be liquidated, especially as that is how the now-industrialised countries tended to view and treat their forests at earlier stages of their development. While many developing countries take a more enlightened view than this, the underlying issues concern equity as much as environment. If developed countries seek to impose limitations on this use by developing countries because of the greenhouse effect or similar concerns, there is a moral case for some compensation on equity grounds. We return to this topic in a later chapter.

Whether in developing or developed countries, publicly-owned native forests almost invariably supply other goods and services besides wood. Our knowledge of the characteristics of these forms of joint production is scant in quantitative terms, yet there is generally a well-founded qualitative appreciation of whether the relationships are moderately or severely competitive or, though this is less frequent, complementary or supplementary (see chapter 1). This in itself is generally sufficient to enable those areas in which wood production might be permitted or restricted to be zoned effectively. However, this zoning is only a first step in dealing with joint production of wood and non-wood goods and services. As we shall see in later chapters, much more needs to be done to deal with the complexities and finer-scale issues of joint production from publicly-owned native forests.

CASE STUDY 5.1 Joint production relationships in ponderosa pine forests.[8]

Beaver Creek watershed in Arizona, USA is the site of an extensive experiment investigating the relationships involved in managing ponderosa pine (*Pinus ponderosa*) jointly for water, wood, scenic quality and wildlife.

Stream gauging and other basic inventory work commenced in 1957. During the period 1969 to 1970, various treatments were carried out on different sub-catchments. These covered a range from 0 per cent to 100 per cent of forest cover removal. The outcomes were monitored and used to make preliminary estimates of the joint production relationships. Scenic quality was evaluated by professional landscape architects subjectively ranking representative photographs of the areas. Other forest uses were measured or estimated using conventional units of measure. The relationships are illustrated for the Brolliar soil type in figure 5.4.

The complexity of the joint production relationships is evident from figure 5.4. Every possible form of relationship is displayed over some range of the various pairs of relationships illustrated. While the sections involving supplementary and complementary relationships can be eliminated to narrow the choice to regions of competitive relationships in the zone between 70 to 50 per cent of cover removed, the choice among these depends critically on the relative prices involved. The preliminary nature of these estimates of relationships should be stressed. Nevertheless, they serve to illustrate the approach.

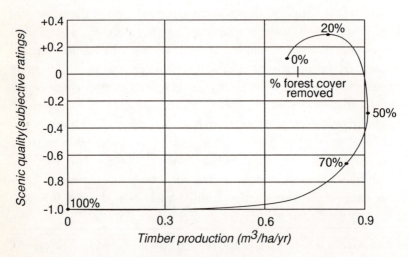

a Joint production of scenic quality and timber.

Figure 5.4 Joint production relationships in ponderosa pine forests, Arizona. Adapted from Brown et al. (1974, Figs 22, 23). The percentage of forest cover removed is relative to a maximum possible removal of 28 m²/ha.

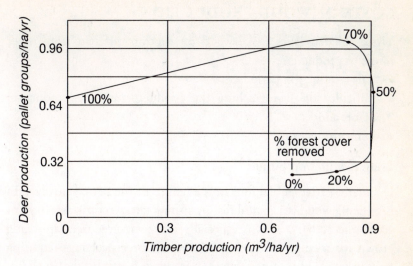

b Joint production of deer and timber.

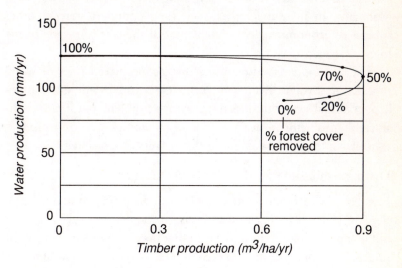

c Joint production of water and timber.

COSTS OF WOOD PRODUCTION

Three characteristics make wood production different from other forms of production:

- long production period,
- difficulty of distinguishing the product from the production plant, and
- joint production.

Long production period

For many native public forests, the length of the period involved in producing wood is a critical distinguishing characteristic as a form of economic activity. This is especially true wherever trees are being grown for medium or large-sized sawlogs or veneer logs requiring trees in excess of 30 cm diameter overbark at breast height. In all but a few cases, the length of the production period (rotation) then extends beyond fifty years and can range from eighty to 120 years. Managing an investment over such long periods is rare in other spheres of economic activity but is part of the stock in trade of the forest manager.

The long rotation also means that economic analyses of wood production involve discounting future costs and revenues. In a purely operational sense this does not present a problem, but it raises considerable problems in dealing with sustainability and joint production issues; these are taken up in the penultimate chapter.

Distinguishing the product from the production plant

Few if any other products are so difficult to distinguish from their production plant. Standing timber, be it sawlogs, pulpwood or both, is difficult to distinguish from the standing tree, both in the sense of the final production outcome and even more so in terms of annual growth.

The final outcome of producing sawlogs, pulpwood and so on is not always apparent from looking at and measuring the standing tree. Many species, especially the less durable hardwoods, suffer from internal defects in addition to those generally more obvious ones associated with branch stubs and occlusions, past damage to the cambium, or malformations of the stem. Internal defects are difficult to ascertain without sampling by boring, cutting wedges or felling, yet they may reduce an otherwise apparently sound tree to the lowest grade of log. Monitoring the external growth of a tree and a forest therefore only provides an initial estimate of the gross change in the volume of various log grades, and may require substantial correction based on extensive sampling or complete enumeration of adjacent coupes when felled. This issue is still further complicated by joint production and varying market conditions.

Joint production

Joint production of various wood, as well as non-wood, products is common because of the range of log grades—such as sawlogs of various sizes and grades, veneer logs, poles, piles and pulpwood—that are produced in all but short-rotation commercial plantations. This characteristic has both good and bad features.

Joint production enables the forest manager to adapt to changes in relative log grade prices by varying the definition of the grades to best advantage. What is a marginal sawlog in today's market may become a pulp log next year. While this is a buffer against the risk of major changes in market conditions, it makes valuing the standing tree and forest even more difficult.

It also makes allocating costs to different products or log grades difficult. At one extreme, the distinctions between jointly produced veneer-logs and sawlogs in one tree or stand may defy separation of growing costs. One is left with arbitrary rules of allocation based on perceived cost-drivers—such as whether the silvicultural regime was primarily oriented to the production of sawlogs or veneer-logs. At the other extreme, pulpwood may often be regarded as a residue from sawlog-driven operations and thus it is readily seen to share none of the costs of growing.[9] As for the joint costs incurred in roading and similar operations, quite arbitrary allocations of costs to particular wood products have to be adopted.

While the problems of allocating costs are not unique to wood production, there are few if any areas of economic activity where they are also associated with long periods of production and difficulties in distinguishing product from production plant; not to mention the added complication of jointly produced non-market goods and services.

BENEFITS OF WOOD PRODUCTION

The benefits of wood production are generally somewhat easier to estimate than the allocation of costs because prices are generally well-established and specific to a product. Much hinges on whether the markets for the wood products concerned are freely competitive, or at least contestable,[10] such that the prices prevailing in the market represent a reasonably close approximation to those that might prevail in a perfectly competitive market and can thus be used to value benefits.

Conditions in many publicly-owned native forests are not always conducive to freely competitive or contestable markets, due in part to exogenous and in part to endogenous factors.

Exogenous factors include corruption or, in more highly developed markets, restrictive practices leading to unfair competition through the exercise of monopoly or monopsony power. Endogenous factors are principally those of location. The nature of publicly-owned native forests is that they are often located in the hinterland away from centres of wood product consumption. Because of the location and economies of scale, the number of processing plants may be limited, providing little scope for competition. The larger and more sophisticated the capital investment, as in the case of bleached kraft pulp mills, the more acute this problem becomes.

Even so, alternative estimates of price can generally be made, using either regional, national or international comparisons of raw material prices delivered at the mill door or through residual value estimates.[11] The latter simply work backwards from competitively established prices for the final or intermediate wood products, deducting the long run average costs of processing, stage by stage, until one is left with a residual value for the standing timber or log. This estimate approximates the derived demand price and can provide a useful proxy for a competitively established price, especially if worked from an intermediate product price and involving relatively simple processing. More complex exercises tend to be less useful because the imprecision attached to estimates of major processing costs is so high as to impute a very wide variance to the residual value estimate.

THE ECONOMICS OF WOOD PRODUCTION

Notwithstanding the difficulties already alluded to, the economics of wood production alone are relatively well established[12] and clear-cut because both costs and benefits can be estimated with reasonable ease. The so-called 'present value' (i.e. discounted value) of the future stream of net revenues can thus be calculated for the various silvicultural options involved. For even-aged systems, where all trees in the stand are of one age, these include the choice of rotation length, the frequency and intensity of thinning, and any other treatments that are consistent with the species and with public ownership. For uneven-aged (or selection) systems, involving a multiplicity of ages and hence sizes in the stand, they include the choice of cutting cycle (interval between successive harvests) and the optimum total volume and distribution of tree-sizes consistent with that choice.

CASE STUDY 5.2: Analysis of rotation length for mountain ash forest.[13]

Mountain ash (*Eucalyptus regnans*) is a species, often the sole tree species, occurring naturally in even-aged stands in southeastern Australia and adapted to regeneration following wildfire.[14] Fire kills existing trees, opens the seed-bearing capsules and thus promotes seed-fall and reduces weed competition during the early stages of seedling establishment. It is also such a tall tree in the mature form (typically over 70 m) that selection systems are difficult to apply because of the difficulties (and the consequent dangers and expense) inherent in attempting to fell single trees or even groups of trees.

In current practice, this forest is generally regenerated by planting nursery-raised seedlings following clearfelling. Assuming that it is grown only for sawlogs without thinning, figure 5.5 shows a graph of the present value of future net revenues for such a stand grown to various rotation lengths, using current prices, costs and an interest rate of 4 per cent. This graph is based on the application of the Faustmann formula[15] such that it assumed that any given rotation length will be followed by an infinite series of identical rotation lengths. This is a device to ensure valid comparison of the economics over an infinite time horizon, rather than comparing the values for disparate hori-

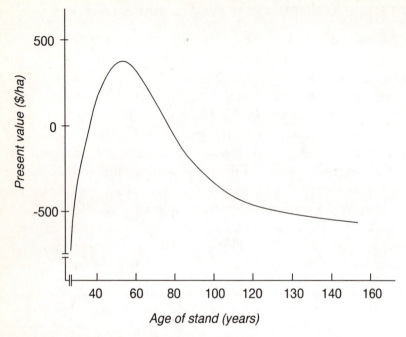

Figure 5.5 Effects of rotation length on present value of wood production in mountain ash forest (Ferguson, 1995).

zons of, say, fifty versus eighty-year rotations, giving the former an unfair disadvantage of a shorter investment horizon.

Figure 5.5 shows a marked peak at about fifty-five years that defines the optimum rotation length for a purely commercial investment. It also shows that if the rotation length is extended much beyond that figure, substantial opportunity costs are incurred relative to the optimum present value.

The choice between even and uneven-aged systems is not always available, as some species cannot be regenerated and grown successfully under uneven-aged systems on some sites. This is especially true of species that occur as fire seres in those zones characterised by frequent wildfires. Nevertheless, where the species can be grown under either system, an economic evaluation of the choice can be made.

Most species grown for sawlogs in publicly-owned native forests exhibit similar characteristics to those shown in the accompanying case study on mountain ash with respect to the optimum length of rotation. The values range from forty to eighty years, very seldom longer, and all exhibit a marked decline in the present value as rotation length is increased beyond these values. These optimum values are seldom adhered to in publicly-owned native forest because of the potentially adverse impact they would have on other forest values. Similarly, those species that are managed under uneven-aged selection systems for sawlog production invariably show levels of stocking and size corresponding to optimum commercial returns that are substantially lower than most managers of publicly-owned native forests are prepared to apply, because of the impacts on other forest values. Therein lies one of the major dilemmas in managing these forests; a matter taken up in subsequent chapters.

part two
resolving conflicts

The management of public forests involves managing widely dispersed stands and areas, often with widely differing biological and spatial characteristics. Because consumers and producers of all forest products are never distributed uniformly in space, there are always imperfections in the operations of markets that stem from this uneven spatial distribution. These imperfections are often exacerbated by the institutional mechanisms of public ownership and supply. Hence a complex overlay of property rights, planning and regulatory processes is needed to moderate these imperfections.

The nested hierarchy of levels at which these rights and processes are most appropriately considered stems in part from the various biological and spatial characteristics of the uses and issues concerning them and in part from the institutional mechanisms for addressing them. The number of levels may vary somewhat from place to place but the following are generally relevant to planning the sustainable management of public forests:

- national planning,
- regional planning, and
- site planning.

INTERNATIONAL AGREEMENTS

Issues such as the greenhouse effect are inherently international in focus because the public consumption impact is or will be global, although that is not to say it is uniform. It is simply not useful to develop and implement control measures unless a sufficient number of nations are prepared to do likewise. The same is true where international trade sanctions are involved, because without collaboration by a substantial majority of producers and consumers, control measures may be ineffective. Conventions, agreements and treaties are the principal mechanisms.

The International Convention for the Protection of World Cultural and Natural Heritage (hereafter, the World Heritage Convention) was adopted by the General Assembly of UNESCO in 1972. It provides for a World Heritage List as an inventory of places of outstanding universal value in terms of their natural or cultural attributes. Properties are nominated for inclusion on the list by national governments that are signatories to the Convention. Following assessment by the World Heritage Committee, an area is inscribed on the World Heritage List if it meets at least one of the several criteria relating to either natural or cultural attributes, all relating in one way or another to an outstanding universal value. Most of the publicly-owned forests listed contain the most important and significant natural habitats where threatened species of animals or plants of outstanding value from the point of view of science or conservation still survive.[1] Most are extensive in area and are protected as national park, wilderness or special reserve. Specific legislation and management arrangements are the responsibility of the national government concerned, but the World Heritage Committee does consider the proposed provisions in making its decision. Recreation may be permitted, in which case listing forests often makes them prominent destinations for eco-tourism. However, some are partly or entirely closed because of the fragility of the communities concerned. Most other commercial or extractive processes are not permitted.

The Convention on Biological Diversity was endorsed at the United Nations Conference on Environment and Development at Rio de Janeiro in 1992 and has met the requirement of ratification by at least thirty signatories. Countries that join the Convention on Biological Diversity shall, among other things, do the following:

- Identify the components of biological diversity important for conservation and sustainable use, and monitor activities which may have an adverse impact on this diversity.
- Develop national strategies, plans or programs for the conservation and sustainable use of biological diversity.
- Establish laws to protect threatened species, develop systems of protected areas to conserve biological diversity, and promote environmentally sound development around these areas.
- Rehabilitate degraded ecosystems and the recovery of threatened species; control risks from organisms modified by biotechnology; and control or prevent the introduction of alien species.

The Convention on International Trade in Endangered Species of Wild Flora and Fauna is another international convention, adopted in 1973. It has prescribed protection for migratory birds and bans on trade for a number of rare or endangered animals, plants, or products thereof, the most notable being ivory. At the 8th Conference of Parties to the Convention, trade in several tropical timber species was proposed to be banned or, for less threatened species, regulated. In the event, the proposals were not accepted by the Conference. Notwithstanding the force of the convention on the parties to it, the process of imposing bans on international trade is generally regarded as being ineffective or discriminatory. It is seen as ineffective because so many countries are either not parties to it or do not enforce the provisions. Many developing countries regard it as discriminatory in banning the export of valuable products sought by affluent developed countries, who earlier squandered their own resources of similar products during their developmental phase.

The International Tropical Timber Agreement[2] was adopted by the General Assembly of the United Nations in 1984 and includes most tropical timber producing and consuming countries of the world. The principal aim may be summarised as involving the achievement of sustainable management of tropical forests by the year 2000.[3] The International Tropical Timber Organisation has been active in examining ways to achieve this aim.

The United Nations Convention on Climate Change was endorsed at the United Nations Conference on Environment and Development at Rio de Janeiro in 1992 and has recently met the requirement of ratification by at least fifty signatories. The convention noted that 'there were still many uncertainties about the timing, magnitude, and regional impacts of climate change but, where there are threats of serious or irreversible damage, lack of full scientific certainty should not be used as a reason for postponing controls'.[4] It undertook to provide help from developed countries to developing countries to deal with the requirements and effects of climate change, through funding, technological assistance, and the provision of environmentally sound technology. It also undertook to provide information and promote sound management of such greenhouse gas sinks as plants, forests and oceans.

Taken individually, these conventions and agreements offer significant hope of improvement in the area concerned. Taken collectively, they provide considerable pressure for the countries concerned to develop comprehensive policies and processes for sustainable management. Yet it is clear that they have as yet failed to achieve the transfer of resources needed to effect sustainable management of publicly-owned forest in developing countries. Implementing sustainable management requires significant additional resources, ranging from an estimated US$0.3 billion per yr[5] to 1.5 billion per yr.[6] Some of those resources are attainable from trade, perhaps to the extent of US$0.1 billion per yr,[7] through better recovery of the economic rents by more realistic pricing of timber, but that still leaves a large unfunded gap.

The United Nations Conference on Environment and Development adopted a non-binding Statement of Principles on Forests[8] that embodies most of the principles discussed in this book for sustainable forest management. The United Nations Commission on Sustainable Development, created to implement the recommendations of the United Nations Conference on Environment and Development, debated the negotiation of a 'Forest Convention', but postponed that measure indefinitely for lack of key political support. It is also by no means clear that the funding for the necessary improvements in policies, processes and practices will be forthcoming.

NATIONAL POLICY

A sovereign nation will need to determine its own laws and policies relating to publicly-owned native forests. It may seek to do so collaboratively with other nations, but the determination remains a sovereign decision to enter into international conventions, agreements or treaties that over-ride or modify national laws and policies.

A national policy aimed at the sustainable use of all publicly-owned forests requires the support of appropriate laws and regulations and the development of land-use planning. At a national level, land-use planning is one of the most fundamental processes towards achieving sustainable forest management. It occupies this central role because it is concerned with identifying the legitimate demands for land to be released for agriculture, urban development and other uses versus that to be retained and developed for forests.

The guidelines established by the International Tropical Timber Organisation[9] state that certain categories of land need to be kept under permanent forest cover to secure their optimal contribution to national development. This is as much an article of faith as a rational and well-founded outcome of logical analysis. It reflects a reaction to an historically long period of continuing decline in forest areas and the underlying fear of tipping the environmental balance too far, as well as a lack of faith in the capacity of private markets to recognise fully all the benefits and costs if they were to be given free reign in land-use decisions. Counter arguments can be made,[10] but the underlying ethic of concern for the forests and their contribution to the environment is so strong that most governments have continued to support the notion of publicly-owned forests.

This highlights one of the anomalies of the planning hierarchy in that the processes become more specific in the underlying bases as one moves down the hierarchy. National planning is often the fuzziest element of the processes, being based largely on the premise that public ownership is a necessity for protection and thus necessitating land-use planning to zone the use of public land appropriately. This explains why ambit claims predominate and the greatest political effort is expended at this level, as the various interest groups seek to influence planning through the political process. National policy-making options are much more difficult to analyse rationally. Lindblom[11] very aptly described this type of decision-making as 'the science of muddling through' or 'disjointed incrementalism', in which a policy-making body is more inclined to proceed by taking a tentative step forward, if only to move two steps backwards, one step sideways, or a further step forward, in the light of the initial outcome.

Regardless of the process of policy and decision-making, the first imperative is to develop appropriate legislation and a supporting framework of policy if public ownership is to be at all effective.

Legislation and policy

As a minimal basis, land-use policy needs to recognise that publicly-owned native forests constitute a common property resource potentially capable of supplying many different goods and services either jointly or separately. The objective is to maximise the social net benefits[12] to be derived from these forests. The purpose of legislation is then to provide for their management and protection in order to achieve this objective. To avoid the problems associated with open access to common property,[13] provision needs to be made for controlling access to the supply of the various goods and services through the definition of forms of tenure and zoning. Because much of the legislation has evolved in stages in developed countries, tenure and zoning tend to be separate operations; the former being dealt with in primary legislation dealing with forests and their reservation and protection, the latter through subsidiary legislation on land-use planning. The detailed forms of legislation and associated regulations need to be consistent with the legal system of the country concerned; discussion of the detail is therefore beyond the scope of this book. Suffice it to say that appropriate legislation and regulations generally do exist, although the provisions for detailed land-use planning are sometimes lacking in some developing countries. Experience in

Australia, New Zealand and elsewhere suggests that the administrative provisions for land-use planning need to be substantially removed from purely regional or local control, because of the susceptibility to influence by local vested interests, and to be open and transparent. This applies with even greater force to the provisions for appeal against land-use planning decisions.

In developing countries especially, attention also needs to be given to allocating property rights to indigenous peoples, who otherwise often lack the political power to ensure that their interests are considered. The allocation of rights may extend to management functions and revenue sharing, but it needs to be done consistently, with identifiable criteria and transparent processes for review of these arrangements to avoid allegations of bias.

However, the principal problem in the developing countries lies less with the legislative provisions for land-use planning and more with implementation. Legislation to protect the permanent forest estate from shifting cultivation and clearing for agriculture is simply not being implemented effectively in some developing countries.[14] Worse still, there is a belief in some countries that the act of promulgating a government decree declaring that forest harvesting must be practised in a sustainable manner actually ensures that that goal is achieved. All experience suggests otherwise. Implementation through moral suasion and enforcement are critical to the achievement of these goals.

Of these two methods, moral suasion is the most important and requires patience and perseverance as well as good communication skills. Winning the support and cooperation of local communities and users is now seen as a more critical and effective means of securing the permanent forest estate than is enforcement, which is principally the method of last resort. Permanent reservation and sustainable management of publicly-owned forests must remain an idle fiction unless (1) forest clearing can be restricted to appropriate and necessary areas, and (2) forest harvesting can be properly regulated on the ground.

According to the International Tropical Timber Organisation, sustainable management is to be achieved by member countries by the year 2000 for all internationally traded tropical timber. Urgent measures must therefore be taken to implement national forest policies in a number of developing countries, especially to curtail illegal or inappropriate clearing for agriculture and shifting cultivation and illegal timber harvesting. This will require close collaboration with agricultural and community welfare agencies and interests. With the exception of occasional showpieces, enforcement in national parks and conservation reserves is especially likely to be neglected because little cash revenue can be derived from them. Where it is imposed, local communities may bear substantial costs in foregoing a cheap supply of subsistence goods for the sake of providing public consumption goods for wealthy urban elites and foreigners.[15] It therefore behoves the international community to take a particular interest in funding the management of these areas and the communities affected.

Legislation and national land-use policy also need to include subsidiary provisions prescribing regional and site-planning measures. These topics are taken up in subsequent chapters.

National forest inventory

An adequate level of knowledge of the area and condition of the permanent forest estate is a prerequisite for reserving a permanent forest estate, land-use planning, determining sustainable levels of production, and estimating any requirements for plantations.

Modern forest inventories of extensive areas involve mapping and then an inventory based on some form of field sampling. The main objective of mapping is to establish a sampling frame of the actual areas of forest by remote sensing,[16] where possible, categorising the forest into strata of different productivity and composition,[17] and identifying and distinguishing areas of shifting cultivation and clearing for agriculture. The technology is changing and rapidly improving in sophistication and resolution. An average cycle of coverage of five years is probably attainable and in keeping with complementary operations.

Forest inventory is complicated by the characteristics of wood production discussed in Chapter 5. These include the joint production of market (wood) and non-market (conservation, water and recreation) goods and services and the inability to distinguish readily between the product (a log in a standing tree) and the production plant (the bole of the tree). Also, the long period required for wood production means holding a very high ratio of standing volume relative to the annual cut. A plantation growing sawlogs on a forty-year rotation involves holding roughly twenty times the volume of the annual cut in standing timber[18] if that cut is to be sustainable. For an even-aged native forest to be managed with a rotation of eighty years, the ratio increases to forty. It becomes even larger for longer rotations, and still larger for selection systems and joint production of other goods and services as well as wood.

The situation in native forests, especially of hardwood species, is often further complicated by the prevalence of internal defect that may render some trees unsuitable for sawlog or other uses. These characteristics make estimating the volumes of standing timber difficult and that of net growth even more so. Inventories are therefore technically demanding and expensive.

A national forest inventory should ideally estimate the present stock of all major forest products, both timber and non-timber, and estimate present and future growth and drain of timber and other renewable products. But this will often be too demanding a task to be accomplished quickly. Indeed, for those countries that are rich in forests relative to present levels of harvesting, the rapid completion of a national forest inventory of the present stock of forest products may be impractical. Field sampling should nevertheless be carried out well ahead of the present boundaries of harvesting and clearing to provide a basis for progressively reserving forest, including identifying conservation, recreation and protection reserves. Field sampling must include flora and fauna identification and inventory, and vegetation and soils mapping. Multi-disciplinary teams are generally the only practicable way of bringing this range of skills together. A cycle of a minimum of five and a maximum of twenty years for full inventory coverage would seem appropriate and practicable.

Major impediments to more rapid progress with national forest inventories include a shortage of staff trained in remote sensing and related technologies and in resources other than timber. Remote sensing now offers a fairly reliable and practicable method of monitoring broad area changes in use, especially the capacity to distinguish between clearing for agriculture, shifting cultivation and selection harvesting.[19] The related technologies, such as geographic information systems, multi-level sampling using conventional aerial and low-level video photography, global positioning systems, field-entry computers and laser-based measuring tools, are advancing so rapidly that their application is in a state of flux. Even those countries that have previously achieved good coverage face daunting problems in updating their inventories.

Land-use planning

Identifying a permanent forest estate and delineating that estate into land-use zones to meet future demands for the various uses is another essential component in achieving sustainable forest management. In its most advanced form, land-use planning involves studies of future demand and supply for all the major land uses and of the net social benefit of those uses. This is seldom achievable in the first cycle of such work but should be pursued in later cycles. The initial priority for land-use planning is to determine and physically mark the zone boundaries of the permanent forest estate.

Most attempts at zoning commence with describing the present characteristics and uses of the forest and its potential use for agriculture and the like, where relevant.

This is not a process for foresters alone. Land-use planning will only be productive if other key departments and disciplines are involved and have a commitment to the outcomes.

Various systems have been developed to assist in this task. One of the more difficult aspects is to identify those areas that might be zoned as reserves representative of the various plant and animal communities present. A coherent network of areas needs to be protected, where ecological processes may proceed with minimal human interference, to assist species preservation and provide opportunities for scientific study. Australia has recently proposed a desirable target of 15 per cent of the areas of forest and woodland types that existed prior to European settlement.[20] The International Union for the Conservation of Nature and Natural Resources[21] recommended that 10 per cent of the remaining distribution of each biome be set as a minimum target for reservation, a much more modest figure. The actual areas chosen by each nation will vary according to the opportunities for reserving relatively undisturbed forest and woodland and the alternative uses for them.

Other areas may need to be protected for special uses, such as recreation or water production. While this zoning is feasible for large areas of diverse and rare forest that might be protected in national parks, wilderness areas, major water catchments, or nature reserves, the scale of national land-use planning does not admit the level of detail needed to obtain a sufficient representation of the more fragmented and smaller communities. This illustrates the reality that an hierarchical system of planning cannot and should not operate as closed compartments: finer detail needs to be pursued at other levels.

Managing the permanent forest estate

The type of organisational structure that should be employed to manage publicly-owned forest has recently become a matter of considerable debate, the current trend being to separate policy and regulatory functions from those of field management. The former functions would be handled by a conventional public service department responsible to a minister or secretary of the government, and the latter by a commercial (generally state-owned) corporation operating under a commercial contract to the policy department but responsible to its own board of directors. Various intermediate forms exist between this new model and the old exclusively public service model.

While these changes undoubtedly have potential benefits, there are some areas of concern for the peculiar functions involved. First, most of the inventory and planning functions of the management entity will be critical for the policy department. Transfer of expertise and data may be effected through contracts let by the department, but the concern is that this department will tend to lack staff with experience in field management and conditions. Once established, the management companies tend to attract a different set of staff from that of the departments by dint of inclination and likely higher salaries. In an area where spatial as well as temporal perspectives are of considerable importance, this is not a trivial concern. Public land management often involves vast tracts of land subject to relatively low levels of travel and use, not to mention poorly funded research and inventory that is necessarily either extensive or very specific in terms of location. All this puts a premium on the experience and observational skills of those who traverse and work in these areas.

INSTITUTIONAL STRUCTURES

Any discussion of the operational aspects of planning needs to be placed in the context of the relevant institutional framework. Traditionally, public ownership has implied planning and management by a single public agency or by two or more agencies, each responsible for a dominant use. That view has been the subject of considerable debate. In part the debate has been driven by political pressures to reduce government spending by hiving off commercial activities and corporatising or privatising them, or contracting out the management of non-commercial activities.

The critical issues for public forest planning and management are concerned with separating planning and management functions, unbundling the property rights associated with this separation, and using corporate entities to manage public forests, subject to a greater or lesser degree of competition.

Separation of planning and management

The term 'planning' is used here in the broadest sense to embrace policy formulation, planning, monitoring and enforcement of those regulatory functions. 'Management' is the implementation of planned activities on the ground.

There has long been a view that planning functions for a government department should be separated from management to avoid the so-called Nader[22] effect of the department being captured by the interests that it exists to regulate. Rothschild[23] brought this to some prominence earlier than Nader in arguing for the separation of research users from research providers in government departments. Such a separation theoretically makes it feasible to ensure greater efficiency in management. Clear goals are given. Managers become responsible for finding the profit-maximising or least-cost solution and thus more accountable than in the typical government department where the political labyrinth of policy formulation, planning, management and monitoring outcomes can sometimes disguise considerable inefficiency and featherbedding. The problem is that this argument assumes that a clear separation of policy and management is possible without engendering additional costs. That is not necessarily easy to accomplish because of difficulties in allocating property rights to public forests.

Unbundling property rights

To take the most extreme view of separation of management functions,[24] suppose all public land was to be zoned into units that were either to be sold or licensed to private or quasi-private management operators. How would that initial zoning be effected?

An enormous amount of information and planning would be required. Twenty-five years of public land-use planning by one land-use planning authority[25] have still not resolved all the conflicting claims for public land. Furthermore, these claims are changing as public demands change. The initial zoning, whether based on area by dominant uses or product-based with overlapping areas for different products, would be extraordinarily complex and controversial. Even if such a zoning could be implemented, there are problems regarding the subsequent planning and management.

Competing corporate entities

To the extent that separation is possible, management by a private or quasi-private entity may achieve greater efficiencies. There is considerable evidence to suggest that the greater clarity of goals and accountability of management in private or quasi-private entities can aid efficiency greatly.[26] However, in the long run, the entry of new management entities and competition or contestability between rival entities are essential to achieving this. This poses problems in the case of public forests.

There is a long history of public utility management in the USA, especially in the supply of water, where natural monopoly conditions made it difficult to exercise that competition. The result was a less-than efficient sector in which the entities achieved the goal of making a prescribed rate of return on investment, but not without considerable inefficiency paid for by the consumer in the form of over-investment, low prices and high operating costs.[27]

In terms of conservation, water production, and many of the recreational uses of public forests, it is difficult to avoid conferring monopoly power on the management entity. In theory, one could fragment the areas of public forest to ensure that competition existed, but this would not be workable in the case of large wilderness areas or catchments or recreational forests, nor would it be compatible with the pleas for a cessation of public forest fragmentation in the interests of preserving biological diversity. Probable economies of scale and scope make larger entities more likely than smaller.

These observations about economies are ones of personal judgment rather than proven characteristics. Few studies have been conducted because of the propensity for these arguments to be driven by ideology.

Of course, there may be scope for some particular and distinctive forests to be managed by private or quasi-private trusts; the example of the Nature Conservancy in the USA often being cited.[28] Even here, a note of caution is warranted to balance the enthusiasm that advocates sometimes invest in this proposition. First, managing the supply of public consumption goods poses problems for such an organisation in that there is no incentive to invest to the appropriate level. Efficient management may therefore not equate with efficient resource allocation and, indeed, one would anticipate under-investment[29] in conservation services, given that the capacity to raise revenues is limited and costs of management substantial. Second, experience with park boards of management suggests that there is a high propensity for them to be captured by vested local interests, notwithstanding the efforts of well-intentioned supporters to the contrary.

Finally, much of the impetus for corporatising public agencies stems from the desire to free them from the shackles of public sector regulations concerning access to capital. However, access to the capital markets carries risks and the penalty for failure to gauge those risks appropriately is the sale of assets and dissolution or bankruptcy of the entity. The exit of inefficient entities from the stage is just as important a form of discipline for encouraging efficiency as the freedom of entry of new competitors. Yet it is difficult to see the latter discipline actually being exercised in relation to state entities managing public forests.

Appropriate institutional structures

Planning and managing public forests pose a unique problem because of the variety of uses with different characteristics involved. It is unlikely that one particular model will suit all situations. Rigid central planning models have been discredited. Completely *laissez faire* models are to be treated with equal suspicion, given the spatial aspects of monopoly power involved and the public consumption good characteristics of some uses.

Certain economies of scale and scope need to be recognised when considering planning and management structures. There are two sorts of these economies: those involving the relationship between planning and management, and those involving their spatial aspects. Planning and management structures also need to utilise the efficiency gains that can be engendered through the use of competitive or contestable processes.

There is considerable logic and potential efficiency gains in having an integrated planning entity responsible for planning all public forests, ranging from national parks to those used for multiple purposes. The planning of reserves and protected areas needs to take into account the overall representation of rare and endangered species and communities, linkages between reserves, and issues of complementary management between adjacent areas of public land zoned for different uses.

There is rather more scope for choice in the arrangements for management. The choices need to be exercised in the light of the desirability of a competitive environment for these entities but are complicated by the hierarchy of planning and management that forms the focus of subsequent discussion. Depending on the economies of scale and scope and the degrees of monopoly power associated with the principal uses, there are valid arguments variously for integrating planning and management, separating planning and management with a single management agency, or separation with multiple and competing management agencies.

7 regional planning

SUSTAINABLE FOREST MANAGEMENT

While sustainability refers to all forest uses and services, not wood alone, it is convenient to deal with sustainable wood production first and then, in the next section, with its integration with other uses.

The conventional view of sustainable wood production involves:

- prescribing and enforcing the amount of wood to be harvested per unit time, such that an approximate balance can be achieved between future growth and the amount cut on a forest management unit; and
- prescribing and enforcing harvesting and subsequent silvicultural measures to ensure adequate regeneration and release of sufficient small stems from competition, and to ameliorate environmental impacts.

The latter is normally prescribed in what has been known variously as a code of forest practice, a silvicultural manual or a field guide and is taken up in the next chapter. The former is normally prescribed in a regional management plan. National plans are too large, with too many sources of error and unknowns, to enable a firm quantitative analysis to be prepared. Administratively, limitations on harvesting need to be calculated and organised on a regional basis to ensure that the spatial and temporal aspects of the issues are adequately recognised.

Sustainability is not a single immutable value for most forest uses. Wood production, for example, can be sustainable over quite a wide range of volumes to be harvested annually. One can harvest lightly and continue to do so, or one can harvest up to the capacity of the site to grow wood and continue to do so. In the past, some foresters assumed that wood production had primacy over other forest uses and hence sought always to achieve this maximum level. That assumption is not tenable on publicly-owned forests today. Much of regional planning activity is designed to elucidate what the appropriate level of harvesting should be.

Data on growth are available for most temperate and sub-tropical forests; many having been managed and measured over a long period. However, few are available for tropical forests, partly because of the relatively recent advent of large-scale utilisation and partly because of the inherent difficulties in obtaining growth data for the many species involved.

Whereas plantations of exotic species in tropical regions may achieve growth of up to 20 m³/ha/yr, Poore and his colleagues[1] used 1.5 m³/ha/yr as a general average for tropical moist forest. This figure applied to the estimated 80 per cent of the total area of tropical forest on which wood production was assumed to be permitted, the remainder being assumed to be held in reserves. These figures may be too optimistic with respect to growth for many tropical countries. For example, Vanclay[2] carried out detailed simulations for North Queensland tropical rainforests and estimated the sustainable level of production to be 0.5 m³/ha/yr over similarly defined areas. Growth rates in native forests are generally much lower than those achieved in plantations because few native forests are managed intensively and most are managed for multiple purposes, involving some trade-off between growth and attributes such as diversity in species, structure and stocking. In all cases, where fuelwood is the major commodity, sustainable levels of production may be substantially higher than those for purely industrial use, because of the capacity to utilise smaller material.

One strategy to relieve the pressure on wood production from native forests is to establish plantations.[3] Such investment ought to be motivated commercially rather than through substantial subsidies that distort the rational pattern of investment.[4] However, the judicious use of minor incentives and provisions in concessions to engage in planting forests has a role as a condition of access to utilising native forests.

Planning for wood production

The current techniques of planning wood production in the long term utilise either mathematical programming or simulation models to schedule the flows of wood to be harvested over time. The need for these scheduling models arises because the forest is never homogeneous. The areas of age classes in even-aged systems are never distributed with sufficient uniformity, nor are the sizes appropriately distributed in uneven-age systems, to enable the harvest to concentrate simply on the oldest age class or the largest trees. Differences in species composition, site productivity and stocking and the occurrence of fires and other natural hazards all combine to make native forest heterogeneous in character. On those areas where it is to be permitted, harvesting may need to be spatially dispersed rather than concentrated and the silvicultural system varied from place to place to minimise the impact on other forest values. The issue is to marry these requirements with the provision of a reasonably stable and secure supply of wood.

Although large and complex in detail,[5] the techniques are simple in principle. A quantitative objective needs to be identified. The maximisation of the present value of net social benefits represents the ideal for a publicly-owned resource. In practice, it turns out that the detail of the objective chosen is less important than the definition of constraints over time, which tend to dominate the realm of choice. Hence the results may differ comparatively little between those found using net social benefit or the present value of future net revenues versus those seeking to maximise the average volume harvested over time. As a matter of principle, however, it would be preferable to use the economic objective, as it provides estimates of the aggregate value of the forest for wood production under the constraints and estimates of the opportunity costs attached to those constraints.[6] If the opportunity cost of a constraint (say) to supply a minimum amount of recreation is extremely high, beyond all reasonable values one might ascribe to recreation, one might wish to reconsider the level of the constraint.

The forest needs to be subdivided into management units that are relatively homogeneous, in the sense that it would be operationally feasible to apply the one silvicultural prescription to all the area within a unit that is available for harvesting. Many hundreds of units, whether coupes, compartments or otherwise defined, are likely to be involved.

The extent of the planning horizon needs to be identified; at times a matter of some controversy. Forest managers have had a tendency to use extraordinarily long planning horizons in the name of considering and ensuring sustainability. However, given the increasing uncertainties attached to far distant planning,[7] it would be more sensible to define a planning horizon to encompass the period over which planning might reasonably be specific as to predictions of change. Beyond that period one has to assume, for want of more confident predictions, entering a steady-state world. The latter fiction does little damage if discounting is being applied because it essentially represents the estimation of the (to be discounted) terminal value of the forest at the end of the planning horizon. In general, a planning horizon of fifty years would seem as much as one could justify. The planning horizon is further sub-divided into discrete planning periods for the purposes of mathematical programming, generally of five or ten years' duration and sometimes commencing with shorter periods and finishing with longer. Conceptually, one-year periods can be used—but increase the computational task accordingly.

The definition of constraints over time then needs to be considered. These will include imposing upper and lower limits on the annual volume to be harvested, consistent with the realities of the market, especially the rational development of dependent industry. For example, in a supply region whose timber resource is expanding as a result of regrowth following a much earlier wildfire, the annual harvest may be capable of progressive annual increase. Yet it may be more sensible to withhold some or all of that progressive increase for a period in order to create a sufficient volume to attract a new industry at a later time. The converse may also be true in the case of reducing the volume harvested, in that progressive reductions may lead to early decay and bankruptcy, whereas a fixed time for closure at or near capacity supply may enable a more orderly retirement of plant and people. These volume constraints are often multiple because different log grades and sizes are involved, with differing future patterns of demand and industrial development.

Many other constraints may be involved: some technical, such as those relating to area, some environmental, such as those discussed in the next section.

The definition of strategies differs for the two major techniques.

In simulation, only one strategy is defined for each management unit and it stipulates the year and silvicultural detail of the operation, be it clearfell, thinning or selection system. One simulation run is therefore a trial of a specific set of strategies, one for each management unit. The outcome is to be compared with another run involving a different set, the process being iterative and limited only by the cost of computing, the data available and the patience of the planner.[8]

In mathematical programming, subsets of alternative strategies (from ten to twenty in number) have to be defined for each management unit and the data relating to each constraint assembled in the linear or mathematical programming matrix for solution. The algorithm chooses the set of strategies across all management units that maximises the objective function. In linear programming models, the optimal solution normally selects only one strategy in the overwhelming majority of management units. Those with more than one may require further (integer) constraints to be imposed, but more often than not are capable of reduction to one strategy by omitting trivial cases.

Whether based on simulation or mathematical programming, these models are not mechanisms that provide a panacea. They are more in the mode of a 'what if' approach, enabling diverse options to be evaluated.[9]

One of the weakness of these models has been the paucity of the linkages to spatial issues, such as the so-called adjacency problem,[10] in which one would generally not wish to harvest adjacent coupes in successive years or periods because of the impact on conservation and perhaps recreational or water values. The development of geographic information systems and their integration with formal techniques of planning[11] may alleviate this problem, with the added benefit of providing a ready display of the spatial outcomes, which is so important for planners and the public alike. Another weakness is the paucity of data enabling the joint production relationships between wood and non-wood goods and services to be incorporated appropriately.

Incorporating other forest uses

Site planning of harvesting operations to protect or reduce the impact on other forest uses is often of greater importance than regional level planning; this is taken up in the next chapter. Nevertheless, wherever other forest uses such as conservation, recreation, or water production can be quantified at the regional level, they should be incorporated in regional planning models. For example, constraints in the model might include setting minimum levels of water production from a major catchment, upper limits on biological oxygen demand in a major river, minimum population size for a wide-ranging animal to ensure viability, or the minimum and maximum levels of forest-based recreation. All are dependent on having data that enable the relevant constraints to be quantified and incorporated in the regional planning model. Ideally, the data for wood and non-wood uses alike should be sufficiently detailed to enable modelling of the risks and uncertainties involved,[12] not just the estimation of mean or median values. However, data collection is expensive because of the extent and variability of native forests. Even for a commercial crop such as wood, data are seldom available for this degree of sophistication in modelling. For rare animal and plant species, data may be available for one or two species, but not for all. Thus while the dynamics of change over time may be capable of modelling simply,[13] they are inevitably incomplete without the capacity to model the dynamics of chance that are so important in managing native forests.

Even so, what benchmarks should be used in gauging unacceptable change given the role of chance? Ecological populations are dynamic: they suffer occasional catastrophes from fire, wind, hail, snow, floods, pest epidemics, drought or volcanic activity under natural conditions, and some communities depend on particular catastrophes for their renewal.[14] For those who argue that native forests should be left in their natural state, this poses a dilemma, because those natural states had changed prior to European invasion or settlement. Other major changes have been imposed since invasion or settlement, including the fragmentation of areas through clearing for agriculture, initially uncontrolled logging, and the initially much increased risk and sometimes intensity of fire.[15] There are no easy answers or absolute benchmarks from the present or past. All we can do is to make choices of what we consider would be a desirable set of states to achieve in terms of species, structures and stockings at the end of the planning horizon, and to adapt that plan to conditions as they arise by regular revision.

Because data are sparse and sometimes incomplete, and the choice of sets of states is partly a matter of personal preference, another process and a set of guiding criteria are needed to supplement the analytical planning process.

Public participation

Reference has already been made to the tendency of the public to distrust paternalistic planning of publicly-owned forest by government agencies. One way of alleviating these concerns is to solicit input and reactions to regional plans from interested members of the public. This is not to argue that the participating public make the planning decisions. Ultimately, the plans need to have regulatory force through approval by statute or regulation and must reflect government and agency policies. However, public input is likely to assist in developing a plan that has broader public consensus and is therefore more readily applied and observed.[16]

The mechanisms for soliciting public input are many; they include advisory committees, public meetings, field days, written submissions on draft plans, and the like.[17] The choice will depend on the local situation and in any event may need to be varied from time to time to avoid entrenched networks developing.

Regional plans need periodic revision every five to ten (at most) years and need to embrace all publicly-owned forests in the defined region. An integrated approach is needed because of the inter-relationships between the management of the different uses of publicly-owned forest within the region. Of course, the regional plan may need to be supplemented by more detailed plans for particular national parks, reserves, catchments or multiple purpose forests, but these need not necessarily have statutory force.

Planning guidelines

Public input should be introduced subject to general policy guidelines concerning the management of publicly-owned forests. While the over-riding objective is to plan so as to maximise net social benefit, this is a counsel of perfection. The data needed to pursue such an objective analytically are often not available. The relative weights to be accorded to efficiency and equity considerations are often unclear. Non-market goods and services pose problems because the prices by which benefits might be gauged are not available. The explicit incorporation of the uncertainties involved is too complex or demanding of data to be feasible. These limitations of data and the analytical models mean that the main objective can at best only be analysed very approximately. It therefore becomes necessary to develop an operational set of policy goals or guidelines that are broadly consistent with the basic objective and substantially capable of reconciliation one with the other. The three main goals are economic viability, environmental sensitivity and sustainability.

Economic viability

A rational agency charged with managing publicly-owned forests should plan and pursue its commercial activities, such as wood production, so as to be economically viable. There is generally no argument for subsidies to wood production, since they would encourage greater investment than is justifiable and so distort the balance between this production and other non-market uses.[18] The one setting in which subsidies may be justified is in developing countries, where fuelwood gathering may constitute an inequitable burden on the poor, and especially on women. However, subsidies are better directed towards encouraging tree-growing on private land to redress the tendency of publicly-owned supply to crowd out the markets.

Economic viability can be assisted by ensuring that the prices obtained for wood reflect their true value through the use of competitive or contestable processes, wherever possible. Economic planning should then aspire to return full costs, including an appropriate return on the capital stock in standing timber. This latter component presents a major assessment problem. The volume of standing timber in a commercial forest managed on an eighty-year rotation is generally at least forty times that of the annual volume harvested. Even if we exclude the standing timber on those areas reserved from harvesting in a native forest managed for multiple purposes, the volumes on the remainder would commonly be very much more relative to the annual harvest. This is because in forests used for multiple purposes habitat and seed trees are retained and there is generally lower stocking of commercial species in the interests of diversity and habitat. The best that can be done, pending complete valuation of market and non-market goods and services in aggregate, is to value in a rather circular manner by calculating the value of standing timber and land based on the discounted value of the future net revenues from timber production. The choice of discount rate then becomes the critical determinant, and this poses problems for wood and non-wood products alike; problems that are discussed in chapter 9.

Nevertheless, there is merit in imposing an economic discipline on investment decisions in roading and similar developments for wood production. Following the principles outlined, investments in roading to access a logging coupe should envisage recouping that investment as well as providing a positive net present value from the sale of wood. If those conditions cannot be met, the harvesting of the area should be cancelled or at the very least postponed until they can be met.

Environmental sensitivity

Environmental sensitivity implies just that; it is not an argument for zero impact on the provision of non-wood goods or services.[19] The guiding principle is that wood production should proceed on those areas available for harvesting up to the point where the additional net benefits derived from it are equal to the additional costs resulting from the impact on the environment. Forest managers need to enumerate the various uses and services provided by the regional forests, hence the need for regional inventories covering all these uses.

Where inventory and related data are insufficient, arbitrary constraints may have to be imposed to provide 'a safe minium level of conservation',[20] where there is a perceived risk of irreversible change beyond that minimum level. Most of the important measures to protect environmental values are, however, applicable to site rather than regional planning and are discussed in chapter 8.

Sustainability

Much has already been said about sustainability. In a pragmatic context for regional planning, considerable emphasis needs to be placed on how the forest will be structured and distributed at the end of the planning horizon. Is the transition into steady state manageable in terms of the changes implied for industry, employment and supply? Are the structures at the end of the planning horizon compatible with the sustainability of animal and plant species, recreation, water production and quality, and other forest products? If not, to what extent should the pattern of wood flows over the planning horizon be modified?

Just as with wood, sustainability of these goods and services is not necessarily about maintaining present levels of supply but about ensuring that irreversible depletion does not occur.

The popular connotation of sustainability is that of constant supply, but this is quite misleading. It is misleading because the state of the economy during the planning horizon is likely to be anything but steady. In the secular sense, the state of the economy is characterised by continuing development and increasing levels of population in most countries. In the cyclical sense, it is characterised by fluctuations in activity that are largely unpredictable. Attempts to impose a steady supply over the planning horizon equate with attempts to stem the tide. Forest management is concerned with the intelligent management of imbalanced forest structures.[21] This requires that wherever possible these changes be addressed in the planning model, not ignored as was the case with the application of a classical 'sustained yield' approach.[22]

Much remains to be said about the concept of sustainability, but this must await completion of our discussion of the pragmatic aspects of three levels of planning.

CODES OF FOREST PRACTICE

The principal distinction between a code of forest practice and silvicultural manuals and field guides is that the former embodies prescriptions for sustaining production of environmental goods and services, as well as those for wood, and includes enforceable sanctions for non-compliance on the part of harvesting or other operators.

Among other things, a code of forest practice should include:

- provision for pre-harvest survey or inspection and the establishment of local reserves for the protection of rare and endangered species and ecosystems or for mass recreation;
- field delineation of:
  –coupe and reserve boundaries and the reservation of nesting trees and other wildlife habitat to sustain fauna and flora populations;
  –the retention of tree species or sizes for regeneration, fruit or nut production, or structural or species balance;
  –the reservation of buffer or filter strips to maintain water quality;
- prescription of harvesting methods and appropriate road design and drainage;
- prescription of amelioration measures to prevent excessive soil disturbance or compaction; and
- sanctions for non-compliance.

Pre-harvest survey and establishment of local reserves

While remote sensing, aerial photography and video photography from helicopters are providing increasingly useful imagery to interpret tree species and structures, there is no substitute for field inspection, especially in old-growth or other forests where rare or endangered species or ecosystems may occur.

One of the most difficult aspects of this work in old-growth forests is that fauna and flora surveys have such different sampling requirements[1] that it is difficult to combine them in the one operation, although there is certainly virtue in having them proceed concurrently to ensure interchange of information between the groups. Fauna surveys are often reliant on trapping, song or call identification, or scat or pad identification. Trapping generally entails restrictions on the time of day—because animals should not be held in traps through the day—and sometimes of season. Some of the other techniques are similarly restricted. Many small animals are relatively specific in habitat, requiring an initial identification of that habitat to provide efficient sampling.

Identifying botanical species can also be highly specialised. While much of the survey work is on an objective sampling basis, often using stratified random or systematic sampling and identification and measurement of tree, shrub and ground-cover species, the identification of rare and endangered species requires acute ocular skills because their occurrence is so irregular. At least this survey work lends itself to integration with surveys for timber production and planning.

The outcome of these surveys is an initial delineation of coupe boundaries on a map of suitable scale, generally not less than 1: 25 000 and preferably larger, together with the delineation of local reserves required to protect flora or fauna or to provide corridors for wildlife movement .

Field delineation of coupe and reserve boundaries

Harvesting teams need to be clear on the boundaries within which they can operate, so delineation of these boundaries in the field is an important and demanding job. The boundaries need to recognise other relevant constraints besides those for flora and fauna protection.

Provision needs to be made for retaining sufficient nesting trees for arboreal species or birds, as well as sufficient seed trees of appropriate species where natural regeneration is to be used. Where selection systems are being employed, special consideration needs to be given to the species and structural balance and stability of the retained trees. Buffer or filter strips of appropriate width need to be provided along permanent watercourses to reduce the probability of soil erosion into the stream itself. Typically, buffer strips of 20 m are prescribed on each side of streams, and 40 m for rivers. No harvesting or vehicle access is permitted in these strips.[2] These strips filter run-off bearing silt or other material and, being the moister sites, often represent the richest habitat for fauna.

Recent research[3] indicates that the proportion of the area closed to harvesting increases almost exponentially with each additional metre of width of the buffer strip. What is lacking in the equation is a clear picture of the consequential change in the expected value of erosion into the stream. Once this is available, the trade-off becomes simple to calculate, remembering that width of the buffer strip should be carried to the point where the additional net revenues from harvesting and timber production is just equal to the additional (expected value) cost of damage to water quality and (if relevant) quantity.

Harvesting methods and road design

Harvesting involves removing forest products, generally wood, from the forest to a point where they can be transported, generally by road, to the processing or consuming centre. For present purposes, it will be convenient to confine the term harvesting to the operations prior to road haulage.

In harvesting, as in other forms of production, the aim is to reduce any negative externalities to the point where the marginal cost of the remaining externality is equal to the marginal cost of reducing that externality. This does not mean total eradication of the negative externality. Total eradication is a thoughtless counsel of apparent perfection that would actually make us worse off because the additional cost of the cure is greater than that of the curse.

Many of the impacts of harvesting also need to be matched against natural disturbances to be placed in perspective. Soil disturbance is never zero in a protected natural forest. Species and structural changes proceed there, as do fires. While these are not arguments against reducing negative externalities, they highlight the fact that total elimination is as unrealistic as it is uneconomic.

The types and characteristics of the wood harvested from the forest vary widely—firewood; posts, poles and piles; pulpwood;

sawlogs. Nevertheless, some general principles apply to all harvesting operations:

- To drag a piece of wood along the ground requires a machine at least three times the weight of the piece of wood to ensure an adequate grip on the ground.[4]
- Lifting wood wholly or partially off the ground dissipates less power in friction but places heavier loads on the machine wheels or tracks or on maintaining an airborne or aerial system.
- Wheeled or tracked machines used to lift or drag the wood can only work at maximum efficiency on a smooth plane. Some earthworks (snig tracks) are generally needed to achieve this.
- To buffer the differences in operating rates between harvesting, log preparation and road haulage, storage areas have to be provided in the immediate vicinity of pick-up points for road haulage. These are called landings and are often the scene of intensive machine activity. Since the advent of log grading and separate marketing of different grades, more log piles are required than previously. This greatly complicates the planning and scheduling of operations and increases the size of landings and associated sites.

The consequences are several. A direct and inescapable consequence is that soil is compacted under areas of intensive machine activity. Soil may also be loosened in the course of earthworks, or otherwise rendered liable to erosion through exposure initiated by machine tracks or wheels. As soil becomes wet, its ability to bear loads and resist machine thrusts becomes less; in some cases, a lot less. The rheological properties may cause the soil to become unstable; for example, soils may suddenly fail after the fifth machine pass.

These effects may lead to increased turbidity and organic matter in streams, as well as in the loss of the more generally fertile topsoil from the areas affected. The latter is a cost to the grower if it leads to lower growth rates in ensuing rotations, but is not strictly an externality.

The impacts of these disturbances are very much dependent on the soil type and on the immediate intensities of rainfall following harvesting. Clinnick and Incerti[5] examined impacts of tractor harvesting on two different soil and forest types. They found that bulk density was not affected significantly on disturbed sites, other than on snig tracks and landings where compaction would be expected to be severe. On one of the soil types, significant changes in hydraulic conductivity from surface sealing were found on disturbed sites, other than snig tracks and landings, but not to a degree that would have rendered the site liable to erosion under high intensity rainfall conditions. Revegetation would quickly redress this effect.

Slope is also a potentially critical determinant of the impact of soil disturbance. Limiting values for tractor harvesting have been developed[6] by transposing the Universal Soil Loss equation[7] to relate slope to acceptable levels of soil loss for a given storm period, rainfall erosivity, soil erodibility, soil cover and conservation practice. Pragmatically, however, the limits for slope are more determined by the operating limits for the tractor concerned and by the soil conditions prevailing at the time. To this end, simple models[8] have been developed for estimating the change in soil moisture resulting from precipitation, given estimates of run-off, interception, evapotranspiration and deep seepage. These seem to provide useful guidelines for closing harvesting operations.

Much greater use is now being made of computers in planning and designing harvesting operations. Access to geographic information systems is becoming widespread and this, together with computer-aided design software for harvesting, offers great advantages in the detail and speed of coupe planning. This is especially true of cable-logging systems where the capacity to design and 'test' alternative layouts with respect to the location of landings and spars is extremely important to the cost and the impact of the operation. In addition, applying expert systems to coupe planning offers opportunities to link in environmental impacts more formally into the planning process.

Much work needs to be done to reduce the pressure of machines on the ground. This is not entirely a machine issue, although the use of wider tyres or tracks is a way of reducing ground pressure. In unpublished field trials,[9] using logging slash as a fibrous reinforcement for the ground has been shown to reduce rutting and shear failure.

Road design and construction has traditionally been handled by harvesting contractors who may be quite efficient commercially but quite unaware of the environmental consequences of their work. Massive cuts and batters with no surface stabilisation are common; bridges, culverts and drains are often temporary or lacking; and road surfaces often lack sufficient hardening. In temperate countries, run-off from roads accounts for much of the sediment load in streams in harvested areas.[10] Field observation suggests that the situation is worse in tropical forests.

Road haulage operations have a different set of potential impacts. Road construction and subsequent drainage are generally recognised as two of the most serious contributors to sedimentation and turbidity in streams; more so than harvesting because harvesting impacts are relatively ephemeral, being soon tempered by revegetation, whereas some roads are permanent or long-lived.

In general, the standards appropriate for feeder roads to provide access to landings are not demanding in terms of earthworks or life-span. Such roads pose no great impact problems unless they are excessively dense.

Permanent or long-lived roads of higher standard do pose a potential problem, whether used for log haulage or for tourist traffic. Untended earthworks may be major sources of point pollution. More important still is the drainage of these roads. Heavily used gravel roads may only constitute 6 per cent of a road network but can produce 71 per cent of the computed sediment.[11] Poor drainage can thus give rise to continuing and major sources of point pollution. The techniques for diverting sediment flows from streams are well-established but they take time and resource to implement and are hence sometimes neglected, to the detriment of the environment. Standards of road design and drainage need to be clearly defined, as do the location of landings and snig tracks.

Coupe supervisors are responsible, on behalf of the forest owner, for developing the on-site detail of marking the trees or boundaries of areas not to be harvested. Given the variation inherent in nature, the planning guidelines represent minimum levels, not absolutes, and a wise and well-trained supervisor will exercise judgment. For example, the width of buffer strips may be extended in a particular coupe in order to preserve riparian vegetation that extends beyond the nominal guideline.

Supervisors are also responsible for communicating the details of the coupe plan to the manager of the harvesting operation and for taking action if the conditions are infringed.

The code of forest practice should stipulate the remedial work that may need to be carried out before the harvesting coupe is 'closed' and considered in a satisfactory condition. Remedial provisions generally include measures such as establishing drainage bars at regular intervals along snig tracks to disperse water flows, ripping compacted areas in preparation for planting or regeneration, dispersing bark piles to reduce fire hazards, and ripping and restoring temporary roads. These operations are prescribed in the harvesting plan given to the harvesting contractor and have to be fulfilled before the operation can be completed.

Sanctions for non-compliance

In some countries and states (e.g. Western Australia), financial penalties are imposed for excessive site disturbance because of snig tracks and landings, or for other breaches of the conditions laid down in the harvesting plan. In others (e.g. Victoria, Australia), the penalties are based on point systems whereby accumulation of sufficient points leads to a period of suspension of the license to operate.

Whatever the form of sanctions, the role of the supervisor is critical to the effective operation of these systems because that person must certify that the harvesting operations in the coupe have been completed to the standards laid down in the code.

SUSTAINABLE MANAGEMENT

The discussion of policies, processes and practices has so far avoided one of the critical unresolved issues concerning the sustainable management of publicly-owned native forests—the selection of the silvicultural system: the bounds within which the rotation for wood production should operate in even-aged systems and those for the levels of stocking and distribution of tree sizes in uneven-aged systems. This is partly because there is incomplete knowledge of the inter-relationships between uses and the valuation of non-market uses, but mostly because the notion of sustainability itself is ill-defined. Hence we turn to these issues in the concluding Part.

issues and conclusions

While the hierarchical approach to planning sustainable management of publicly-owned native forests described in the preceding Part provides a robust and pragmatic means of making progress towards that goal, there are many unresolved issues and difficulties. The matter of sustainability itself is by far the most important of these; it is taken up in the next chapter. Some of the other issues are described briefly in the concluding chapter.

9 sustainability

SUSTAINABLE DEVELOPMENT

Sustainable forest management has been seen by many as a logical extension of the principle of sustainable development as defined by the Bruntland Commission[1]—that current needs are to be met as fully as possible while ensuring that the life opportunities of future generations are undiminished relative to the present. Such a definition leaves considerable scope for manoeuvre and little specific guidance operationally. Yet there has been comparatively little attention given to its refinement.[2] Indeed, there is sometimes a disinclination to pursue it, as witness a statement by the leaders of one international workshop on sustainability of small islands in which it was agreed that sustainability should not be discussed because everyone intuitively knew what it was.[3] In this sense, sustainable development (or sustainable forest management) might well be regarded as a mandala—a ritualistic symbol or icon of some desired but ill-defined future.

Some of the most severe critics have argued that sustainable development is a contradiction in terms. If so, it follows that sustainable forest management must be likewise. Others argue that, with respect to timber production, sustainable management is simply not achievable. The study of tropical forests by Poore et al.[4] concluded that only 800 000 ha of the total area can be regarded as being under sustainable management. It highlights that much has to be done to achieve or even approach sustainable management in tropical forests and reinforces the cynicism of these critics.

Among the critics, there is often a deep distrust of any commercial exploitation of publicly-owned native forests and of its consequences for the conservation of biodiversity and other values. These protagonists see sustainable forest management as a myth designed to justify continued timber production.

SUSTAINABILITY

If for many of the critics sustainable management is a myth, for many of the enthusiasts it is a mandala. The underlying problem leading to these diametrically opposed views is that neither the objective nor the criteria for the underlying conceptions of sustainable forest management have been given sufficient definition to enable one to be sure that the protagonists are referring to the same things. Before sustainable management becomes a mandate through international convention or agreement and if it is to represent more than a passing fashion, it will need to be defined more precisely.

One of the major issues to be addressed is the gulf that is apparent between economic theory and some of the concerns that underlie the notion of sustainability.

Neoclassical economics

The neoclassical theory sees capital as embracing both manufactured goods and natural resources, both having a high degree of substitutability, such that Dasgupta and Heal[5] argued even in the absence of any technological progress, exhaustible resources do not pose a fundamental problem. They go on to qualify that statement to refer only to allocative efficiency and note that it may not be palatable in other respects. Nevertheless, in caricature, the neoclassical view is that sustainability is not an issue because economic output can be maintained indefinitely through substitution; a process greatly assisted by technological progress.[6]

In practice, the neoclassical approach to sustainability simply applies cost-benefit analysis to choose those investments capable of making potential Pareto improvements—i.e. of yielding positive net benefits such that the surplus potentially enables compensation of anyone made worse off by the investment, thereby potentially leaving no-one worse off and some better off as a result of the investment.

The forest rotation problem

There are many ways in which the apparent failure to account for sustainability in neoclassical capital theory can be expressed. One of the simplest and most pervasive is the forest rotation problem.

In perfectly competitive markets, the efficient use of natural resource capital, such as harvesting standing timber in a forest, is governed by well-established theorems derived from capital theory. The so-called Faustmann[7] solution requires that an even-aged stand[8] be harvested when the present value of net revenues from the present and all future such rotations reaches a maximum. An example was given in figure 5.5.

Yet wherever this solution has been mooted for or applied to publicly-owned native forests, the reaction of managers and the public has been adverse because of the impact on the supply of other forest uses, the rotation being much shorter than seems desirable. There are numerous examples of actions taken in public forests that are simply not justifiable in terms of the return to investment in wood production. These include increases in the rotation length in even-aged stands and lighter selection of stems to be removed and longer cutting cycles in uneven-aged forests. As figure 5.5 shows, to defer the clearfelling beyond about eighty years incurs a large opportunity cost if timber production is to be pursued.

One response to this problem is to include the net benefits of the various other uses or externalities associated with that forest in the analysis of the efficient use of capital. Leaving aside the practical difficulties of valuation using shadow prices for the unpriced goods and services, this is a sensible step to take in developing a more complete model of joint production.[9]

Where publicly-owned forests with special characteristics occur, it may well be that their value for providing non-wood goods and services alone exceeds their value in producing timber jointly with those other uses. Under these circumstances, harvesting should not be undertaken.[10]

Yet it is difficult to extend this argument to embrace all publicly-owned forests, many of which lack any special characteristics or scarcity values (see chapter 4). In these latter forests, the annual streams of net benefits from other uses are more likely to be quite modest. These are the forests in which producing timber jointly with those other uses has a clear rationale. However, reliance on conventional economic analysis would lead to joint production using relatively short rotations that forego the net benefits from non-wood services associated with older ages. Even if the future streams of net benefits from non-wood uses were to be included, the discounting process would render those small streams irrelevant or negligible beyond sixty years or so. Thus, even if conservation and other unpriced goods and services are valued together with timber in joint production, the forest rotation problem is still likely to emerge[11] because, for any real rate of interest about or above 4 per cent, the contributions to present value of future streams of net revenues beyond about sixty years diminish rapidly when discounted.

At one time, it was conjectured[12] that the optimum rotation length for wood production alone might be longer than the results of the deterministic analyses in figure 5.5 would indicate if risk was taken into account. A higher variance attached to the present values of shorter rotations might lead a risk-averse decision-maker to opt for a longer rotation, notwithstanding a lower mean for the present value for the longer rotation. However, the empirical substance of this argument has not been borne out—indeed the limited evidence suggests that the variance does not change sufficiently to support this hypothesis.[13] In any event, a public official charged with decisions on rotation length that constitute a small fraction of national income should adopt a risk-neutral approach, not a risk-averse one.[14] Thus, on both grounds, risk offers no answer to the rotation length problem.

Figure 5.5 is used purely to illustrate the nature of the problem that flows from the peakedness of most rotation length analyses. The magnitude and location of the peak in time will vary widely with species, site, and price–size relationships, among other things.

Natural capital

Several economists[15] have argued that the fault lies in the all-embracing definition of capital, which presumes that all elements of capital are potentially substitutable, one for the other. Accordingly, they separate manufactured from natural capital, such that all manufactured capital is assumed to be substitutable, but not natural capital.

Unfortunately, 'natural capital' lacks any clear definition. Is a forest of a species that has been silviculturally treated or genetically improved to give higher yields a natural capital? The growing stock and products may be nearly indistinguishable, other than genetically, from a naturally regenerated crop. Natural capital is identified with natural resources and includes a diverse array; some with flow, some with stock characteristics; some renewable, some not; and some with durable products, some not. Definition is thus difficult. The operational use of a generic but fuzzy term such as natural capital is fraught with difficulty.

Nevertheless, the proponents of this view argue that maintaining the natural capital stock, not just capital in its entirety, is needed for sustainability, so that the next generation will have at least the same stock of capital, including natural capital, as the last. Drawing the distinction between natural and other forms of capital is motivated by the inability of economic markets to recognise environmental values adequately where unpriced and public goods are involved.

In a recent work, Pearce and Warford[16] argue that it is also necessary to distinguish between various forms of natural capital, especially those forms that have a low degree of substitutability and that might therefore be deemed critical capital. They posit different relative values being accorded to different forms of capital, in recognition of the different degrees of substitutability. Yet this view is qualified by the statement that:

> ...lack of substitutability, by itself, does not necessarily imply high value. That is, we would expect a tropical forest containing a rich store of biodiversity to have a higher value than, say, machinery...because markets rarely exist for the life-support functions of natural environments.

At the very least, this highlights the arbitrary character of their valuation of natural capital. But it also highlights the issue of valuation in any natural capital approach. If valued from the future flow of net benefits discounted to a present value according to the conventional processes of valuation, little or no weight is given to the flows enjoyed by generations seventy years hence because of the effect of discounting over such a period at conventional rates of interest.

Finally, why should a country or region[17] rich in particular natural resources not diminish its natural capital in that resource, if it is to the net benefit of present and future generations. Yet the natural capital approach says otherwise or, alternatively, suggests that substitution of another form of 'natural capital' for forests would be acceptable. Historically, most developed countries have taken the path of initially diminishing their forest capital. In many cases, one could concede that the process may have gone too far, in ignorance of the consequences, but that is the wisdom of hindsight. The precedents are not lost on those developing countries who still have a substantial forest resource. They naturally perceive current attempts by developed counties to restrict the level of the cut in developing countries as a quite unfair measure, especially when much of the greenhouse effect problem stems from the developed countries themselves.

Inter-generational equity

Norgaard and Howarth[18] re-examined the capital theory approach and developed a macro-type capital theory model incorporating overlapping generations that enables the inter-generational aspects to be encompassed. Put briefly, their argument is that this problem is simply a matter of equity, which the economic theory cannot determine. Inter-generational equity implies that redistributive transfers may have to be made to redress the welfare of later generations relative to the present. Given knowledge of those measures, the neo-classical model is then capable of determining the most efficient allocation of resources.

The problem with this approach and others that deal with overlapping and multiple generations[19] is that essentially they ascribe a say in the investment decision to future as well as present generations. The social welfare or utility functions incorporate separate utility functions for the different generations, rather than the social welfare or utility functions of the present generation including the utility the present generation derives from consumption by later generations.

While perhaps morally admirable, this approach does not and cannot accord with the realities of decision-making. It is the set of present generations who must make the choice. Hopefully, they will be mindful of moral exhortations concerning future generations, wildlife preservation and the like, but they and they alone can make the decision. Thus if inter-generational equity is at issue, it has to be resolved by those present and voting or investing.

Ethics

If inter-generational equity is a matter of ethics, are there any other ways in which the ethical choice can be integrated with the economic theory? Kohn[20] recently advanced a novel treatment concerning the existence value of wildlife, based on the Bergson–Tintner–Samuelson framework of a theory of social interest, separate and distinct from private interest. The individual might be said to have a split personality: one half guided by purely individual motives and utility, which may nevertheless derive some personal utility from the conservation of the wildlife in question; and another social persona that derives utility from the benefits to others and future generations that will flow from that conservation.

As Stevens and his colleagues[21] point out, this is but one of at least two groups of theories that recognise ethical behaviour as involving something potentially distinct from the neoclassical model of self-interest and individual utility. While this type of approach integrates the conceptual bases of ethical and private choice, its application to problems rests on evaluating the parameters of ethical choice[22] and social utility, one approach for which is taken up in the following section.

Social rate of time preference

Arndt[23] points out that the 'discount rate is a device used by economists to quantify their subjective judgement of the social rate of time preference, the relative weight given to present and future income streams.' The social rate of time preference is thus intended to reflect social preferences for consumption (or income) over time.

The evidence suggests that the real (i.e. deflated) social rate of time preference lies in the region of 4 to 6 per cent.[24] These are rates that are broadly compatible with what we observe in the market place after taking out the effects of inflation, risk and any opportunity costs relevant to the differences in capital management between the private and public sectors.

The social rate of time preference is the geometric product of the marginal rates of time preference for successive time periods (years).

Marginal rates appear to be constant up to the maximum term of investment transactions of, say, thirty years. However, few loans or transactions extend longer and most are much shorter, so the picture may be different beyond that horizon.

Suppose we conducted a survey designed to elucidate the marginal rate of time preference exhibited by the set of present generations towards those variously four and five or six and seven generations removed from today (say, 100 plus years). The question to be posed is 'would you prefer an additional $100 of consumption to accrue to the fourth generation or the fifth generation'.[25] My hypothesis is that people are increasingly unwilling to express a marked preference between generations so far distant. The time period is so great, the uncertainties so large and the choices so abstract that such discrimination, be it between the fourth and fifth or the seventh and eighth generations, seems unlikely. This leads me to a hypothesis that the marginal rate of time preference of the present generation is substantially lower for consumption by successive generations from the fourth generations onwards, say eighty years.

If this hypothesis is accepted, then logically there must also exist a transition period during which the marginal rate changes progressively from the present stable level exhibited by markets up to (say) thirty years ahead to a lower value at that time ahead (eighty years) when our preparedness to discriminate between future generations is much less, say a marginal rate of time preference of 2 per cent.[26] The three phases are illustrated in figure 9.1.

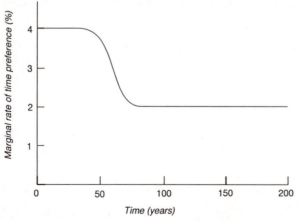

Figure 9.1 Hypothetical trend in marginal rates of time preference over time.

If this three-phase approach to marginal time preference is accepted, then it follows that the value of the social rate of time preference, which is a geometric product of the marginal values up to the year in question, declines progressively from fifty years onwards, as illustrated in figure 9.2. At the limit, the social rate of time preference approaches 2 per cent as time approaches infinity.

These, of course, are ethical hypotheses about marginal rates of time preference. They reflect our perceptions about inter-generational equity, but they involve a variable that can be applied directly to the subsequent analysis of efficiency. The same results could be reproduced by choice of particular levels of inter-generational transfers,[27] but this requires that the present generation assess the preferred choices for future generations, which is more difficult and less direct.

The effects of such a recognition of inter-generational equity are several. First, the treatment is compatible with capital theory and cost-benefit analysis. Second, it gives greater weight to net benefits derived by future generations than was the case and those weights may be capable of empirical confirmation, at least for the zero marginal time preference phase. Whether one treats this as a matter of ethics or of efficiency is unimportant because it is readily integrated in the

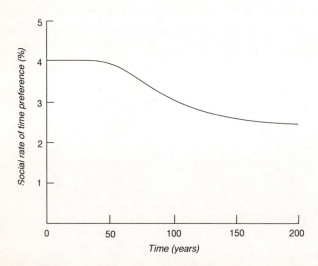

Figure 9.2 Social rate of time preference over time based on hypothetical trend in Fig. 9.1.

economic analysis. A wide range of existing property rights that reflect cultural values and history are taken as given in most economic analyses and yet reflect underlying ethical issues.[28] This is not to argue that economic models are capable of solving ethical issues, merely that the distinction is not clear-cut. Where ethical values can be integrated acceptably into the one model, there is every reason to do so.

The approach is not without problems. It raises the possibility of multiple optima in terms of rotation lengths because of the changes in the effective social rate of time preference with time. But these conditions are not solely the product of this approach. They frequently arise because of the complex surfaces that describe the inter-relationships between joint uses, and the spotty distribution of imprecise data.

More importantly, it leads to solutions that may not be consistent when re-evaluated by some later generation. However, that is both their prerogative and duty, because of the uncertainties that cloud all projections and predictions. There is always a need to re-evaluate strategies in the light of changing conditions.

SUSTAINABLE FOREST MANAGEMENT

While this exposition may assist our understanding of what sustainability entails, it does not answer all the operational problems relating to sustainable forest management. Sustainable forest management will not be rendered precise and analytically operable through the definition of sustainability based on inter-generational equity and its measurement through the social rate of time preference. Uncertainties attached to the actual marginal rates of time preference, probable existence of multiple optima, inadequate data on the relationships between uses and their prices, and imperfectly defined property rights mean that analytical perfection will continue to elude us, although that is not to minimise the need for improvement in all those respects.

Under these circumstances, there is a strong case for adopting a conservative approach in planning the level of cut, the selection of the silvicultural systems, and the rotation length in even-aged systems or the analogue of levels of stocking and tree-size distribution in uneven-aged systems. The basis of the case rests first on the possible economic contribution of the non-wood values, some of which may become more substantial with longer rotations, and the notion of sustainability discussed in the previous section. There is also evidence in land-use decisions in the public arena which suggest that the public at large place a high value on non-wood uses. It may be, of course, that these decisions were ill-founded and irrational, but the pervasiveness over time and location suggests otherwise. The difficulty is where to draw the line, because that line should logically vary with the particular circumstances.

A country or state with a large public forest resource relative to its population has far more scope to make both adequate provision for protecting representative ecosystems and for varying degrees of commercial wood production in publicly-owned forests. Where the use of the forest is almost exclusively for commercial wood production, the rationale for holding it in public ownership becomes tenuous and the choice of silvicultural system and rotation purely commercial. Where other forest uses jointly assume importance in those areas on which wood production can be practised, then the level of cut, choices of system and rotation need to be deliberately modified.

Planning the level of cut

Mention has been made of the use of computer-based planning models. The question arising from the preceding review of sustainability is how to incorporate those matters in the regional planning process.

One could apply the proposed social rate of time reference function to a model with an infinite or very long planning horizon, say 100 years, consistent with the long rotations that seem appropriate for joint production values. However, the specification of constraints becomes an empty exercise beyond about fifty years because the uncertainties attached to forecasts are so great.[29] Thus it seems more appropriate to use a more restricted planning horizon of, say, fifty years and to focus attention on a comparison between the spatial and age (or size-class distribution, in the case of uneven-aged forests) structures of the forest now and those in fifty years time, modifying the latter in the light of professional and public perceptions of the likely conservation, recreation and water production status of the forest as it seems likely to exist at the end of fifty years. This provides a more tangible goal for the interested parties than does a projection to 100 years under conditions so uncertain that the projection has little meaning.

Of course, the resulting plan needs to be reviewed every five to ten years. Contractual arrangements for the supply of wood or other products may impose limitations on the capacity to change radically, unless through compensation by buying out all or part of the contract, but those are the realities of management for any commercial uses, including recreation and perhaps water. Generally there will be sufficient flexibility to make significant if not radical changes. Some degree of flexibility is essential if only to cater for natural hazards such as fire, wind, hail, snow and pest damage; not to mention the inevitable cyclical fluctuations in the commercial markets themselves.

Silvicultural strategies

Most forestry agencies are already using a range of silvicultural strategies that are substantially more conservative than those indicated by purely commercial considerations in relation to wood production. However, there is scope to differentiate more between forests of different site productivity, where better sites might well have a range of rotation lengths shorter than those for poorer sites and include thinning and fertilising options.[30] Pushing some of the regrowth forest more rapidly towards a mature structure by such means can have potential conservation and recreation advantages in offsetting risks posed to remaining old-growth forest by fire.

There needs to be a greater variation in the choice of silvicultural system and size of coupe, where possible. Single tree selection systems are generally not practicable in very high forest, over 70 m, because of occupational safety considerations in felling. Some species also require clearfelling on some sites if satisfactory regeneration is to be obtained. Where clearfelling is used, there is sometimes a rational preference on the part of wildlife and conservation experts for somewhat larger coupes rather than very small coupes, in order to avoid spreading the disturbance by vehicles, roading and noise over a wider area. This greater level of detail requires the use of 1:25 000 maps as a basis, and preferably 1:12 500, if sufficient detail is to be incorporated into regional planning.

In applying codes of practice, more emphasis needs to be placed on varying the detail of the silviculture to suit the non-wood as well as the wood production uses of the coupe—appropriate measures for retaining habitat for endangered species, feathering boundaries in areas seen from recreational roads and viewing points, identifying erodible soils and treating them and so on.

Sustainable forest management must be based on a pragmatic set of processes, at least for the foreseeable future. Hence the emphasis on a hierarchy of policy and planning measures. The national, regional and site measures are not totally discrete and independent, nor are they complete or unambiguous, but they are appropriate to the consideration of management of any publicly-owned forests or woodlands. This is so whether or not wood production is involved, for the same issues arise with respect to joint production of the non-wood uses.

There are no easy solutions. If we accept the premise that public ownership is necessary to protect the multiple values of these forests, we must also accept that governments must establish these or similar processes for planning and regulating forest management. The challenge is to make these processes work effectively at a time when public faith in governmental planning is low.

10 conclusions

INTRODUCTION

Global concern regarding the sustainable management of forests is gathering momentum, following the United Nations Conference on Environment and Development at Rio de Janeiro in 1992. Publicly-owned native forests are the principal focus of conflict and concern.

The reasons for this focus are several. First, these are the forests which because of their species composition and location potentially have multiple uses, including conservation, recreation, water production, 'minor' forest products, and wood production. Of course, not each and every use is necessarily available from all areas. Second, being publicly-owned, these forests are the subject of competing claims: from entrepreneurs and industries interested in commercial products; from non-government (predominantly urban) organisations interested in conservation and recreation; and from rural peoples who live in or around those forests and who derive sustenance or livelihood from them or suffer or risk loss from feral animals, noxious weeds or wildfire. Third, these forests represent by far the largest proportion of the world's forests. Yet their global area has been declining fairly rapidly—a matter of special concern in relation to tropical forests that has assumed greater prominence in view of possible global warming due to greenhouse effects.

One of the outcomes of the United Nations Conference was a Statement of Principles on Forests that may form the basis of further negotiations.[1] Though nebulous in present form, it heralds one step in a process that may lead to a binding agreement on sustainable forest management.

The International Union for the Conservation of Nature and Natural Resources[2] and most international conservation groups recognise that it is neither practicable nor desirable to advocate protection (in the sense of banning harvesting and other extractive activities) of all publicly-owned native forests, and thus support the move for sustainable management. Any attempts to prevent all extractive uses of publicly-owned native forests face the twin problems of the management cost for conservation mentioned in chapter 3 and, especially in developing countries, the threat of illegal use on a scale and of a character that is far worse in its impact on the natural environment than any planned extractive use. Examples of the problems include illegal fuelwood harvesting involving repeated cutting without regard for regeneration, illegal harvesting of high-valued saw- or veneer-log species without regard for regeneration or damage to residual stems, illegal grazing by cattle or goats, and fire and shifting cultivation following illegal harvesting

The International Tropical Timber Agreement of 1989 set in train the goal that, by the year 2000, all tropical timber traded internationally should come from forests that are sustainably managed. Many tropical producing nations are now arguing that the Agreement, due for renewal in 1995, should be expanded to encompass sustainable management for both temperate countries as well. Some are also arguing that the FAO Tropical Forest Action Plan and other FAO Forestry Division activities should be combined with this enterprise to create a single and much larger world authority for overseeing forests and forestry.[3] The United Nations Commission on Sustainable Development is due to report in 1996 and its report may add impetus to negotiating an agreement on the sustainable management of forests, albeit through its interim decision to postpone that negotiation indefinitely.

Much of the emphasis on sustainable management refers to those forests used for timber production, because timber production is the largest and most widespread commercial use with the greatest impact on other joint products and uses. The extent and intensity of harvesting, whether in terms of the extent of clearfelling in even-aged systems or of the proportion of the trees selected in uneven-aged systems, have been a continuing source of debate in relation to their impact on other uses. Nevertheless, there is general agreement that sustainable management should not be seen as pertaining to the regulation of harvesting alone—it concerns all uses and all native forests.[4]

INSTITUTIONAL MANAGEMENT

The International Tropical Timber Organisation guidelines for sustainable forest management place considerable stress on securing a permanent forest estate. While this is clearly a necessary step, it will be of little avail if the institutional framework to follow the guidelines is ineffective, as it is in many developing countries.

Dove[5] refers to this as the 'law and order' issue. Laws and regulations to protect forests are of little value if corruption is so significant or staffing levels or salaries are so low as to negate enforcement.[6] Both legal and administrative enforceability are necessary conditions for sustainable management.

However, laws, regulations and staff are necessary but not sufficient conditions to ensure that a permanent forest estate can be maintained. In countries where population pressures on the forest resource are extreme, as in many parts of India, villagers must be brought to recognise that conserving the forest is in their own long-term interests and that encroachment will continue to erode the resource. Well-meaning attempts by non-government organisations frequently assume that winning the hearts and minds of villagers is simple, neglecting the deep divisions between caste and socioeconomic status within villages.[7] These divisions make the poor and landless relatively more dependent on the communal forest resource for subsistence, but nonetheless eager to increase their land holding through encroachment, as others do. Medium and higher income villagers, having more resources and being more powerful in both a political and an economic sense, also pursue encroachment, often on a more substantial scale than the poor. There is often no enduring sense of a community ethic favouring conservation of the forest resource and censuring private appropriation of public land. The tendency is to see public land as privately appropriable by stealth or corruption, partly for personal gain and partly because of an antipathy towards government officials in their attempt to enforce restrictions on the use of that land—attempts that are often distorted by corruption or over-zealous or inept interpretation of the law and regulations. For these countries, massive and sustained programs of education through the political institutions, schools, media, and government agencies are an essential complement to the role of laws, regulations and staffing in developing sustainable management.

While the development of private investment in tree and forest resources in these countries has much to recommend it at both village and industrial levels, it will not provide the range of uses that are or can be provided by public forests, especially the public consumption goods like conservation of flora and fauna. Rural villages with rapidly increasing populations are likely to continue their private appropriation of public land, thus endangering the conservation of flora and fauna, because appropriation requires few resources. Thus, these countries must not only develop the law and the means of enforcement; they must also develop a respect for the law through an appreciation of and support for the ethic of public property rights. This is a very major and difficult extension of the argument for planting more trees for wood and non-wood material goods. It generally requires some form of participation in community management of the public forests if private appropriation is to be defeated. The utility of international campaigns and aid would, in my view, be much greater if directed more along these lines and less along single-issue, high-profile campaigns directed to particular elements or areas of flora and fauna.

The problem is not confined to the legal and administrative institutions—markets are often 'corrupted' or less effective than they should be because of the undue exercise of monopoly power or other imperfections.[8] At the very least, market contestability[9] needs to be developed wherever possible.

ALLOCATION OF PROPERTY RIGHTS

Securing a permanent forest estate involves land-use planning based on prior forest inventory of the various present and potential uses, their production rates and stocks. Land-use planning requires delineation of the areas that are intended to remain under forest cover and the prescription of those uses to be excluded from particular areas of forest.

A coherent network of areas need to be protected where ecological processes may proceed with minimal human interference to assist species preservation and provide opportunities for scientific study. Other areas may need to be protected for special uses, such as recreation or water production. Securing the permanent estate is equally important but is now seen as much a matter of winning the support and co-operation of local communities and users as one of enforcement.

In developing countries especially, attention needs to be given to allocating property rights to indigenous peoples who otherwise often lack the political power to ensure that their interests are considered. The allocation of rights may extend to management functions and revenue sharing but needs to be done consistently, with identifiable criteria.

In recent times, the argument has been made that many of the problems relating to managing publicly-owned forests as a common property might best be resolved by conferring the rights to management and sometimes even ownership to local communities, raising questions as to the future role of public ownership of forests in the traditional sense.

Private ownership may work for particular areas where the net benefits can be substantially appropriated by the owners without the use of monopoly power. Where substantial appropriation is not possible and management rests largely on the altruism of the owners, the situation is unlikely to be stable in the long run as new owners seek to extinguish unwanted burdens. Private ownership is least appropriate where a substantial supply of public consumption goods and services is involved, as in the case of most subsistence goods and conservation services, much recreation, and some water supply services.

As with strictly private ownership, communal management or ownership is no panacea. In developing countries, experience with communal management or ownership suggests that it often tends to be inequitable in sharing the net benefits among members of the community, with the wealthy securing rents and the poor or ill-educated being squeezed even harder in relation to subsistence uses of the forest. There are many examples of community boards of management in developed countries being captured by particular interest groups, be they real estate, grazing, conservation or timber interests, to the disproportionate benefit of that particular group. Even where equity within the local community is not an issue, access to potential net benefits for the wider regional or national public may be denied by local communities. For example, it is difficult to conceive of local communities in developing countries generally being receptive to the argument about the need for conservation of biological diversity, because those net benefits are only cogent for the broader community.

In short, we are stuck with the age-old argument of balancing local against broader community interests. Surrendering all management or ownership rights to local communities is unlikely to provide a stable or sustainable system, unless accompanied by checks and balances. Any system involving extensive checks and balances will itself be dynamic, because barriers inevitably raise opportunities for particular groups or individuals to exercise monopoly power and capture rents. Such a system will therefore need modifying from time to time, sometimes shifting the balance of decision-making towards the regional or national level and sometimes in the reverse direction.

By the same token, however, public ownership should not be taken to imply one particular and permanent regimen of property rights at the local community or any other level. What is appropriate for the better rainfall areas may not work in areas of marginal rainfall in Kenya, any more than the same measures will work for high forest compared with semi-arid woodlands in Australia, or southern compared with northern forests in Sweden. Periodic review and change is needed to take account of changing external and internal conditions.

SUPPLY MANAGEMENT

The steps involved in supply management are well recognised and often assume the greatest importance in the literature, notwithstanding the importance of the other elements in this section.

In brief, sustainable management on the supply side requires integrated national, regional and site planning of all uses on a temporal and spatial basis. The operational criteria are those of economic viability, environmental sensitivity, sustainability and public participation.

Planning models need to work directly from geographic information systems and to deal with spatial adjacency problems,[10] where recent operations on one coupe may constrain or prevent operations on adjacent coupes for a period. However, the spatial resolution of such models is inevitably limited and a safety net is needed to encompass both regional and site planning.

That safety net can be provided through an enforceable code of forest practice,[11] which, in combination with pre-harvesting site planning and subsequent monitoring, ensures that operations are handled in an environmentally sensitive manner, with provisions for stream buffer zones, wildlife corridors, special reserves for rare and endangered species, retention of habitat trees, amelioration of potential pollution or erosion problems and disincentives for transgressions of the code.

DEMAND MANAGEMENT

Demand management has become increasingly important with the recognition that it can legitimately be influenced by governments in various ways to relieve undue pressure on a particular resource or use.[12]

Demand management is principally achieved through incentives to research and development to encourage better utilisation, further processing and recycling. If these measures are insufficient, incentives or disincentives can be introduced to encourage the use of substitutes. Finally, if short-term crises arise, advertising and moral suasion can be used.

PUBLIC PARTICIPATION

With all the preceding steps, the role of the public needs to be borne in mind. The public is collectively both the owner of the basic production plant and consumer of the goods and services supplied. Because many of these goods and services are unpriced, and some have public good characteristics, it is essential that the public be involved and consulted as much as possible in the development of plans. Purely technocratic planning is neither acceptable nor conducive to better solutions.

Critics will rightly point out that the distinction between public and private ownership is not as clear as the introduction and general tenor of this book would suggest, as witness the long history of public access to walking and berry-picking in privately-owned forest in Sweden and similar public rights of use on privately owned land in other countries. Property rights include the right to exclude others from use of or benefit from the area of land, the right to transfer that right to others, and any conditions placed on the use or uses. Historically, the concept of private property was associated with conferring the first two rights subject to very few (but seldom no) conditions. However, the sharpness of the distinction between publicly-owned and privately-owned land has been blurred as rights of access to particular uses have been granted on the one and conditions have been imposed on the other. Public participation often provides a litmus test for granting these rights and imposing conditions of use.

CONCLUSIONS

Sustainable management of publicly-owned native forests will increasingly be an international issue. An international agreement requiring sustainable forest management seems likely. Yet sustainability itself lacks clear definition and, unless that definition is forthcoming, sustainable management may lose credibility. This book has argued that sustainability is about inter-generational equity, and has advanced a conceptual basis for dealing with it in principle and a pragmatic basis for applying it in the context of sustainable forest management.

In any event, sustainable management requires more than just the definition of sustainability. It requires effective institutions, whether legal, administrative or market-based. It requires the rational allocation of property rights through land-use planning and other measures. It requires the management of supply through regional management plans and site planning, complemented by an enforceable code of forest practice. It requires demand management to ease population and other demand pressures on publicly-owned native forests. It needs public participation because of the dual role that the public plays as both consumers and owners of the means of production.

Those who support sustainable forest management should recognise that it will not be achieved quickly or inexpensively. Sustainable forest management by the year 2000 for all tropical timber traded internationally, the goal set by the International Tropical Timber Organisation, is unrealistic and unattainable even for many member countries. There are cogent questions as to the legitimacy of confining this agreement and goal to tropical timber. The challenge is to strike an appropriate balance between the scale of the problem and the resources and time available, and to structure appropriate institutional mechanisms to administer the process.

notes

1 introduction

1 The definition of what constitutes 'comparatively little' intervention is open to debate, as is the apparent emphasis on a forest as a collection of tree species without reference to other components of flora or fauna. Operationally, however, these are the primary forms of intervention and emphasis for forests that are widely regarded by the public at large as being 'natural' by comparison with most plantations or forests made by humans. Scientists and conservation groups may, and often do, have quite different views.

2 The dichotomy between native forests and plantations of native species is not always as sharp as this statement would imply. As plantations of native species reach older ages, beyond fifty years or so, they frequently take on many of the properties of apparent 'naturalness' associated with native forests and are often mistaken for native forests by the public at large.

3 Franklin et al. (1981) provided one of the first attempts to define the ecological characteristics of 'old-growth' Douglas fir forest in North America and greatly influenced the development of definitions elsewhere. However, as Burgman (unpublished) points out, subsequent studies suggest that notwithstanding some similarities, many differences exist in the ecological attributes that characterise old-growth in different forest types. Hence any attempt at a general ecologically-based definition could be potentially misleading. For the purposes of classification, simpler definitions are often used, such as that of Dyne (1991): 'forest that has less than 50 percent of the primary canopy as mature and senescent trees as measured by Aerial Photographic Interpretation'. Such a definition is implicit in the sentence used in the text but has the disadvantage of implying a Clementsian view of succession that is not intended.

4 Again, this is from the viewpoint of the public at large. Scientists and other discerning observers will almost certainly be able to identify differences. As my colleague, M. A. Burgman (pers. comm.) has put it, 'the inability to distinguish may depend on how hard you look'.

5 Dasgupta et al. (1972) and Little and Mirrless (1969) reflect somewhat different approaches, although technically identical in outcome. Hansen (1978) provides a good review of the procedures favoured by the World Bank at present.

6 McNeely (1988).

7 Unhappily, this term is often contracted to 'public good'.

8 The distinction between tradeable and non-tradeable commodities is now a standard procedure for cost-benefit analysis. See Hansen (1978).

9 See Dasgupta et al. (1972) and Hansen (1978) regarding exchange distortions.

10 ibid.

11 Mitchell and Carson (1989).

12 Hotelling (1938), Clawson and Knetsch (1966), Ferguson and Greig (1972).

13 Sinden (1974); Sinden and Worrell (1979).

14 The classic reference is Carlson (1965).

15 Carlson (1965) uses the term 'technically independent' but this lends itself to mis-interpretation.

16 Carlson (1965) omits reference to this, even though it is well recognised by Marshall (1966) and others.

17 Gregory (1955) was probably first responsible for correcting the prevailing view in forestry that multiple use was automatically superior.

2 conservation

1 The readings in Elliot and Gare (1983) provide an introduction to the ethical positions involved.

2 McNeeley (1988) and others divide these further into existence, option and bequest values but the distinctions are not critical to the analysis of the benefits involved.

3 Sinden and Worrell (1979) note that the amount raised substantially exceeded the purchase price, indicating a clear social valuation that the land was worth more in conserving this species than in agriculture.

4 Based on the most recent draft definitions (Version 2.2) under discussion by the International Union for the Conservation of Nature and Natural Resources. These draw heavily on the work of Mace and Lande (1991) and represent a major change from the earlier definitions in

Groombridge (1992), which contained no explicit reference to risk assessment. These and the earlier definitions form the basic of the periodic revisions of the so-called Red Lists of which Lucas and Synge (1978) and the International Union for Conservation of Nature and Natural Resources (1990) are two examples. Individual countries and states within countries have developed their own modifications of these definitions.

5 Mooney (1993) provides some estimates of the contribution of South germ plasm to the value of North wheat and maize production but the basis of these estimates is unclear, especially with respect to the contribution of 'wild' genes.

6 Graham et al. (1990).

7 MacArthur and Wilson (1967).

8 Boeckeln and Gotelli (1984).

9 Harris (1984), in particular, draws heavily on island biogeography theory in dealing with the management of public forests in the Pacific Northwest of the United States.

10 Mace and Lande (1991), Lindenmayer et al. (1991), Burgman et al. (1992).

11 Case study based on Lindenmayer and Possingham (1994).

12 Cunningham (1960). The present author was engaged in some of the fieldwork in that study. A clear distinction needs to be made between those areas burnt in fires of high intensity in the 1939 wildfire and those (e.g. in the north of Ada Block) that were burnt by fires of lower intensity. Very few trees and presumably possums survived in the former, in which salvage harvesting was carried out over the next five years or so. Many or most trees and presumably possums survived the fire in the latter, which were harvested in the mid-1950s, variously using tractor or winch logging.

13 Shaffer (1987).

14 Small populations are subject to loss of rare alleles and thus tend to have lower probability of long-term survival, as well as being more prone to extinction from random environmental fluctuations. Nevertheless, small populations subject to stressful conditions may lose heterozygocity more slowly than those in benign environments (Lesica and Allendorf 1991).

15 See Reznicek as cited in Lesica and Allendorf (1991).

16 Botkin et al. (1972) developed the Jabowa model and showed it to be reasonably accurate and realistic.

17 Noble and Slatyer (1981) set the scene for models that incorporate gap and spatial effects.

18 Margules and Austin (1991) provide a useful review of cost-effective methods of survey, data analysis and reserve design.

19 Remote sensing and geographic information systems have greatly improved the capacity for analyses of this kind. Remote sensing has enabled classification of areas of tropical forest that were previously intractable to stratification by aerial photography, in a fraction of the time required (Vanclay 1994). Geographic information systems provide both powerful tools for database management and useful techniques for analysis of the data.

20 Kirkpatrick and Brown (1991).

21 Joint Scientific Committee (1990).

22 International Union for the Conservation of Nature and Natural Resources et al. (1991).

23 This and subsequent material draws heavily on an unpublished report by Burgman and Ferguson (1995) and on an unpublished paper by Burgman (pers. comm.).

24 Whitmore (1990).

25 Ibid.

26 Jaeger (1988).

27 This is a somewhat more modest figure than the 465 million ha estimated by Sedjo (1989) requiring an investment of US$186 to 372 billion.

28 Much of this reforestation will involve naturally regenerating or reseeding of areas of logged and degraded land, rather than plantation establishment. According to R.N. Byron (pers. comm.) of the Centre for International Forestry Research at Bogor, some 200 million ha of such lands await rehabilitation in South America.

29 Woodwell (1992), National Academy of Sciences et al. (1992).

30 See Pimental in Botkin and Talbot (1992)

31 Whitmore (1990).

32 Dickinson and Henderson-Sellars (1988).

33 The balance between the supply from native forests and that from intensive cultivation is unclear because statistics on the former are scant. However, the balance certainly appears to be shifting towards intensive cultivation.

34 Ruitenbeek (1990, 1992).

35 Botkin and Talbot (1992).

36 Ibid.

37 Gershon (1992) reported a somewhat controversial agreement between Merck & Co Ltd and InBio, a Costa Rican agency, to allow screening of a collection of some 500 000 plant specimens in return for US$1

million per year for two years and 5 per cent royalty on any commercial products developed from them.

38 Brown (1993).

39 Balick and Mendelsohn (1992).

40 Ibid.

41 Balick et al. (1994).

42 Balick and Mendelsohn (1992).

43 Richards (1993).

44 As Richards (1993) points out, expectations of the potential contribution of non-timber forest products in providing long-term sustainably managed systems have been badly overstated. Among other things, concerns are held that intensive cultivation of the products concerned is inevitable because of the difficulty of maintaining supply and quality of supply from extensive operations in natural forests. While this may be true of the areas close to development and with a vigorous cash economy, it is not necessarily so of those forest dwellers at the extensive margin of economic production.

45 Resource Assessment Commission (1991, 1992).

46 Arrow et al. (1993). See also Chapter 3.

47 As with most public consumption goods, it is possible to envisage special conditions where this is not true; for example, for persons entering an enclosed, sealed space.

48 Kanowski et al. (1992).

49 O'Regan and Bhati (1990). The costs of management for conservation include those of fire protection, pest control, visitor services, monitoring, inventory, planning, and protection against illegal activities. R.N Byron (pers. comm.) of the Centre for International Forestry Research at Bogor argues that the costs of managing for conservation could be much lower in developing countries if carried out by local communities. Even if that were the case, the relativity of costs to those for areas managed for wood production might not be dissimilar to those for Australia, because the principle of local community involvement in management is equally applicable to areas managed for wood production in developing countries. However, my own experience suggests that where local community management does effect a reduction in the costs of forest management, much of the gain is taken up by the additional expenditures required on public education and monitoring, whether through government or non-government organisations, in order to maintain a community ethic of good stewardship and to avoid corruption and private appropriation of public goods.

50 Concise summaries of these issues are sparse. Some for insect habitats

are given in the study by Smith (1994) concerning the impact of theTransamazon road corridor. Some of the diseases are introduced there and many of the health problems are more related to poverty and disturbance of the environment than to the natural environment itself.

3 recreation

1 Douglass (1982).
2 Crowe (1966), Litton (1968) and McHarg (1969) pioneered the modern approach to landscape management. Most forest management agencies have their own manuals to assist field managers. US Department of Agriculture Forest Service (1973, 1974) provides a comprehensive treatment of landscape planning and forest recreation. Forestry Commission, Tasmania (1983) and Leonard and Hammond (1984) provide other examples.
3 Case study based on Litton (1968).
4 See Manning (1985) for useful summary of these problems.
5 See Lucas (1990).
6 See Manning (1985) who summarised the early studies of 1950s and 1960s based on the ORRRC studies and later research.
7 Nash (1973).
8 Manning (1985).
9 See Litton (1968).
10 Australian House of Representatives Standing Committee on Environment and Conservation (1977).
11 Nash (1973), Stankey (1973).
12 See Shaw and Williams (1994) for a comprehensive treatment of tourism.
13 Hotelling (1938) developed this method: See Clawson and Knetsch (1966) for a simple illustration.
14 See Kalter and Gosse (1969) for examples and Cicchetti (1973) for a general treatment.
15 Cummings et al. (1984), Mitchell and Carson (1989).
16 See Mitchell and Carson (1989), the review commissioned by National Oceanic and Atmospheric Administration in the USA (Arrow et al. 1993) and subsequent reviews by Portney (1994), Hanemann (1994) and Diamond and Hausman (1994).
17 Case study based on Dixon and Sherman (1991).
18 Ibid. Note that this estimate was derived from users. Non-users were not surveyed.
19 This may be attractive in the short term, enabling the higher consumers'

surplus of the overseas tourists to be tapped to a greater degree but leads
to long term problems of discrimination and resentment on one or both
sides relating to services received or not provided. Discounts for school
children and families on an otherwise uniform fee would seem a better
long-term strategy.

20 Unlike the estimates elsewhere in this case study, this is a personal estimate
by the author based on the total volume available and some knowledge of
conditions and potential net revenues. It should be noted that all harvesting of native forests in Thailand is now banned, following the occurrence
of major landslides that were subsequently attributed, probably incorrect-
ly, to forest harvesting.

4 water

1 Satterlund and Adams (1992)
2 This treatment is a considerable simplification of the basic elements of
forest hydrology developed in more detail by Brooks (1991) and
Satterlund and Adams (1992).
3 See Stadtmuller (1987) for a detailed discussion. Persson (1974) indicates a total area of about 500 000 km^2.
4 Bruijnzeel (1990).
5 See Satterlund and Adams (1992) for a detailed account of the variable
source area concept. In tropical climates, rainfall intensity may be so
high that the infiltration capacity of very porous soils is overwhelmed.
6 See Kuczera (1985) and O'Shaughnessy and Jayasuriya (1991).
7 See Satterlund and Adams (1992, p34).
8 Satterlund and Adams (1992).
9 Boada (1988).
10 Browder (1988).
11 Leitch et al. (1983).
12 Glick (1991). Sole attribution to humans may be inaccurate because
some animals native to the area are vectors.
13 Stiglitz (1986).
14 Ibid.
15 These alternative means are referred to as the 'backstops', a term attributed to Nordhaus (1973) pioneering study of energy resources by
Fisher (1981).
16 Clarke (1994).
17 This case study is based on a study by Ferguson (1995) using data in
Read et al. (1992, 1994).
18 Kuczera (1985).

19 This model was developed by Clarke (1994). The specific estimates are given in Ferguson (1995).

5 wood

1 Food and Agriculture Organisation (1994).
2 Chambers et al. (1989).
3 Based on forecasts in Food and Agriculture Organisation (1991).
4 Food and Agriculture Organisation (1994)
5 Based on forecasts in Food and Agriculture Organisation (1991).
6 Food and Agriculture Organisation (1994).
7 Based on forecasts in Food and Agriculture Organisation (1991).
8 Brown et al. (1974).
9 Leslie (1985).
10 Baumol (1982).
11 See Repetto and Gilliss (1988) for examples of international comparisons and O'Regan and Bhati (1991) for residual value and other comparisons.
12 Gregory (1972); Leuschner (1984).
13 This case study is based on Ferguson (forthcoming). The economic analysis is based on a single price for sawlogs, regardless of size, and may tend to exaggerate the peak in figure 5.5 and shorten the optimum rotation. Nevertheless, the general character of the observations made here are quite robust with respect to changes in price and species.
14 Attiwill (1994a,b).
15 Faustmann (1995) [1849].

6 national planning

1 United Nations Educational, Scientific and Cultural Organisation (1987).
2 United Nations (1984).
3 International Tropical Timber Organisation (1991b).
4 Keating (1993).
5 Ferguson and Munoz-Reyes Navarro (1992).
6 Keating (1993).
7 Ferguson and Munoz-Reyes Navarro (1992).
8 Keating (1993).
9 International Tropical Timber Organisation (1991a).
10 For example, see Stroup and Baden (1983).
11 Lindblom (1959).
12 As always, there are both efficiency and equity dimensions to this criterion and hence the difficulty of analysis.

13 For examples, see Chambers et al. (1989).

14 There are numerous developing countries where this is patently the case (Poore et al. 1989).

15 I am indebted to R.N. Byron for reminding me of this point.

16 In this context, aerial photography can be regarded as another form of remote sensing.

17 For examples, see Vanclay (1994) and Silviconsult (1990).

18 This follows from the 'normal forest' model exemplified by the von Mantel formula for determining the allowable cut (Davis and Johnson 1987), where the allowable cut is equal to twice the volume of growing stock in a normal forest, divided by the rotation length. Transposing terms enables the cited result to be derived.

19 Silviconsult (1990).

20 Department of Primary Industries and Energy and Department of Environment, Sport and Territories (1995).

21 McNeely (1993).

22 Named after Ralph Nader, the crusading environmentalist who first brought the effect to public prominence.

23 Rothschild (1971).

24 Anderson (1991), Stroup and Baden (1983), Moran et al. (1991).

25 The Land Conservation Council of Victoria is now engaged in the second round of public land-use planning since its inception in 1970.

26 Kikeri et al. (1992).

27 Hirshleifer et al. (1960).

28 Tietenberg (1992).

29 Ibid.

7 regional planning

1 Poore et al. (1989).

2 Vanclay (1994).

3 International Tropical Timber Organisation (1991c).

4 The subsidies currently granted to pulp companies in Indonesia are excessive and are promoting rapid conversion to plantations that are far from fully proven in terms of their economic viability. If they do not prove viable or attain the present predictions of yield, the onus for supply may fall on the remaining native forest.

5 Dykstra (1984) and Davis and Johnson (1987) provide excellent treatments of the detail.

6 The opportunity costs of a binding constraint will be apparent through the value of the dual price provided by the linear programming solution.

7   See Ascher (1978) for some vivid examples of the errors attached to forecasts of population, gross domestic product and other national statistics in the USA some five to thirty years ahead.

8   See Davis and Johnson (1987) for binary search and dynamic programming techniques. These to some degree bridge the differences between the simulation and linear programming techniques described above.

9   Hof (1993).

10 Burgman et al. (1994)

11 O'Hara et al. (1984).

12 Burgman et al. (1993).

13 Unfortunately, case studies of these changes are voluminous and defeat any attempt to present a summary here, but two examples on widely differing regional scales include Department of Conservation and Environment, Victoria (1992) and Forest Ecosystem Management Team (1993).

14 Fire seres such as the mountain ash forests of Victoria and Tasmania and the Douglas fir forest of the Pacific Northwest of North America are examples. While there are some analogies, the argument presented by Attiwill (1994a,b) that harvesting and regeneration activities in mountain ash forests emulate those natural disturbances seems a little extreme. Nevertheless, his point that disturbances are naturally occurring processes is important because it highlights the dilemma for those that argue that the environment should be maintained as it is or was.

15 With early settlement in Australia, traditional and relatively regular hunting fires lit by Aborigines declined. Escapes from agricultural burning under very severe conditions then became the predominant source of fire. These were of much greater intensity than those of the Aborigines (Pyne 1991).

16 Cubbage et al. (1993).

17 Ibid.

18 See Repetto and Gillis (1988) for examples of these problems.

19 As with environmental pollution, it is simply not rational to pursue zero impact or zero pollution goals as there is a trade-off between the benefits and the costs (Baumol and Oates 1979).

20 Ciriacy-Wantrup (1952).

21 Clutter et al.(1983).

22 Ibid.

8 site planning

1 This chapter draws heavily on Ferguson and Bren (1991).
2 See Myers et al. (1983) for a comprehensive discussion of the various survey methods.
3 Clinnick (1985).
4 Bren (1995).
5 Caterpillar Tractor Company (1988).
6 Clinnick and Incerti (1985).
7 Clinnick et al. (1984).
8 Wischmeier and Smith (1978).
9 Cornish (1981).
10 C. Humphries.(pers. comm.). Forest Harvesting Manager, Australian Newsprint Mills, Albury, April, 1991.
11 Campbell and Doeg (1985).
12 Reid and Dunne (1984).

9 sustainability

1 World Commission on Environment and Development (1987: p. 8).
2 The work of the Ecologically Sustainable Development Working Groups (1991) sponsored by the Australian Government achieved some refinements in terms of specificity to particular resources such as forests but little on the conceptual approach.
3 Jasisuriya (1992).
4 Poore et al. (1990: p. 196).
5 Dasgupta and Heal (1979: p. 196–7).
6 Empirical investigations of this hypothesis have yielded various results. The early work of Barnett and Morse (1963) generally supported it, but later studies by Slade (1982) and Hall and Hall (1984) cast doubt on those conclusions. Norgaard (1992) goes further and argues that such studies are logically fallacious, being founded on the very hypothesis that they purport to be analysing.
7 The original solution is due to Faustmann (1995) [1849]. See also Gaffney (1957) and Samuelson (1995)[1976].
8 In practice, regional public forests generally consist of a set of stands of different ages and site productivities with a degree of monopoly power due to location and scale. Net benefits become the sum of consumers' and producer's surpluses and various constraints have to be applied to the optimisation to smooth fluctuations in supply and in the demands for

inputs. Stand level analysis illustrates the principles and problems more simply and vividly.

9 See chapter 1 and Gregory (1955).

10 Clarke (1994) provides a useful analytical approach for timber and water applicable to other non-wood uses producing annual streams of net benefits.

11 Hyde and Newman (1992) argue that the emphasis on joint products from most forests is often unwarranted and thus, presumably, that the neoclassical solution for timber production should generally stand. However, recent events concerning the spotted owl controversy in US National Forests suggest that public concerns about unpriced goods and services are significant because they involve very large opportunity costs in terms of the resulting constraints on wood production.

12 Dowdle (1965).

13 Campbell (1974) provides an empirical analysis of the distributions for mountain ash.

14 Arrow and Lind (1971).

15 Hartwick (1977) is credited with the so-called Hartwick rule that sustainability requires that the pure profits from the use of exhaustible resources be invested in renewable natural capital, thereby maintaining the stock of natural capital. Hartwick (1978), Pearce and Turner (1990), Constanza and Daly (1992) and Common and Perrings (1992) provide variations on the theme.

16 Pearce and Warford (1993: p. 53).

17 There are, of course, arguments such as global warming or a concern for biodiversity that might cause other countries to offer compensation to reduce or stop this practice—see Sandler (1993).

18 Howarth and Norgaard (1990), Howarth (1991), Norgaard (1992), Norgaard and Howarth (1992).

19 Bellinger (1991).

20 Kohn (1993).

21 Stevens et al. (1993).

22 Other ethical theories include that of natural rights, which maintains that wildlife (sometimes taken to include plants) have a right to exist independent of any value to humans (Stevens et al., 1993; Nash, 1989). This is an ethical proposition that has some currency but not general acceptance. Like most ethical theories, it seems incapable of integration with the neoclassical model unless as a moral exhortation prior to instituting a Rawlsian 'veil of ignorance' in which a utopian social welfare function is evolved free from present biases. There is no evidence of this happening.

23 Arndt (1993: p. 657).

24 Ferguson and Reilly (1976), Lind (1990).

25 Being consumption, the amount is not transferable from the recipient to the other generation.

26 None of the values used in this hypothesis have any empirical basis and should be regarded as being illustrative only. Several authors, including Solow (1992), have argued that the effective social rate of time preference may be small and lower than the market rate of interest. Some, including Rothenburg (1993), have drawn a clear distinction between discount rates to be applied to normal commercial transactions, including those across generations, and those involving partly or wholly non-commercial transactions and public consumption goods across generations.

27 As Norgaard (1992) demonstrates.

28 Landlord/tenant laws and regulations, cooling-off periods on hire purchase and other transactions, and other aspects of commercial law are examples.

29 See Ascher (1978) for many examples of the imprecision of forecasts beyond ten to thirty years.

30 See Raison et al. (1995)

10 conclusions

1 Keating (1993). Reference has already been made to the decision of the United Nations Commission on Sustainable Development deferring negotiation of a 'Forests Convention' indefinitely because of a lack of key political support. Nevertheless, I would be surprised if this matter was not raised again in the next five years as the extent of clearing and uncontrolled logging of tropical forest becomes even more apparent.

2 Holdgate (1993).

3 Brand et al. (1993).

4 It also extends to private forests and plantations but the objectives of management are so different from those for publicly-owned native forests that a distinction needs to be made in the requirements imposed on them.

5 Dove (1993).

6 All countries are sensitive about allegations of this kind and it is therefore worth remembering that the same problems often beset developed countries at an earlier stage of their development and still do, to some extent, though on a lesser scale.

7 Nadkarni et al. (1989).

8 Repetto and Gillis (1988). Imperfect markets in forestry in developing

countries are common, in the author's experience, as indeed they were in presently developed countries at earlier stages of development.

9 Contestability implies freedom of entry and exit.

10 O'Hara et al. (1989) illustrate an approach that overcomes this limitation of linear programming models commonly in use.

11 There are many examples, among them that of the Forestry Commission Tasmania (1987) and Department of Conservation, Forests and Lands, Victoria (1988) codes in Australia.

12 Demand management techniques were first developed and applied to the energy industry but have since been widely applied to water.

references

Anderson, T. L. 1991, *The market process and environmental amenities*, Centre for Independent Studies Occasional Papers 34, Centre for Independent Studies, St Leonards, Australia.

Arndt, H. W. 1993, Sustainable development and the discount rate, *Economic Development and Cultural Change*, 41 (3), 651–61.

Arrow, K. J. and Lind, R. C. 1971, Uncertainty and the evaluation of public investment, *American Economic Review*, 60, 364–78.

Arrow, K., Solow, R., Leamer, E., Portney, P., Radner, R. and Schuman H. 1993, Report of the National Oceanic and Atmospheric Administration, Panel on Contingent Valuation, 58 *Federal Register* 4601.

Ascher, W. 1978, *Forecasting: An appraisal for policy-makers and planners*, Johns Hopkins University Press, Baltimore.

Attiwill, P. M. 1994a, The disturbance of forest ecosystems: the ecological basis for conservative management, *Forest Ecology and Management*, 63, 247–300.

Attiwill, P. M. 1994b, Ecological disturbance and the conservative management of eucalypt forests in Australia, *Forest Ecology and Management*, 63, 301–46.

Australian House of Representatives Standing Committee on Environment and Conservation 1977, *Off-road vehicles: impact on the Australian environment*, Third Report (A. B. C. Wilson, chairman), AGPS, Canberra.

Balick, M. J. and Mendelsohn, R. 1992, Assessing the economic value of traditional medicines from tropical rain forests, *Conservation Biology*, 6, 128–30.

Balick, M. J., Arvingo, R. and Romero, L. 1994, The development of an ethnobiomedical forest reserve in Belize: its role in the preservation of biological and cultural diversity, *Conservation Biology*, 8 (1), 316–17.

Barnett, H. J. and Morse, C. 1963, *Scarcity and growth: The economics of resource availability,* Johns Hopkins Press, Baltimore.

Baumol, W. J. 1982, Contestable markets: An uprising in the theory of industry structure, *American Economic Review,* 71 (1), 1–15.

Baumol, W. J. and Oates, W. E. 1979, *Economics, environmental policy, and the quality of life,* Prentice-Hall, Englewood Cliffs, New Jersey.

Bellinger, W. K. 1991, Multigenerational value: modifying the modified discounting method, *Project Appraisal,* 6 (2), 101–8.

Boada, E. L. 1988, Incentive policies and forest use in the Philippines. In *Public policies and the misuse of the forest resource* (eds R. Repetto and M. Gillis), pp. 165–98, Cambridge University Press, Cambridge.

Boeckeln, W. J. and Gotelli, N. J. 1984, Island biogeographic theory and conservation practice: species-area or specious-area relationship? *Biological Conservation,* 2 (1), 63–80.

Botkin, D. B. and Talbot, L. M. 1992, Biological diversity and forests. In *Managing the world's forests: looking for balance between conservation and development* (ed. N. P. Sharma), pp 47–74, Kendall/Hunt Publishing Company, Dubuque, Iowa.

Botkin, D. B., Janak, J. F. and Wallis, J. R. 1972, Some ecological consequences of a computer model of forest growth, *Journal of Ecology,* 60 (3), 849–72.

Brand, D. G., Roberts, R. W. and Kemp, R. 1993, International initiatives to achieve sustainable management of forests: Canada's model forests, the Commonwealth Forestry Initiative, and the development assistance community, *Commonwealth Forestry Review,* 72 (4), 297–302.

Bren, L. J. 1995, Aspects of the geometry of riparian zones and their relation to logging management, *Forest Ecology and Management,* 75, 1–10

Brooks, K. N. 1991, *Hydrology and the management of watersheds,* Iowa State University Press, Ames.

Browder, J. O. 1988, Public policy and deforestation in Brazil. In *Public policies and the misuse of the forest resource* (eds R. Repetto and M. Gillis), pp. 247–83, Cambridge University Press, Cambridge.

Brown, A. G. 1993, Global use of Australian forest genetic resources. In *Australasian forestry and the global environment* (eds R. N. Thwaites and B. J. Schaumberg), Proceedings of the Institute of Foresters Australia 15th Biennial Conference, pp. 179–86, The Institute of Foresters of Australia, Canberra.

Brown, H. E., Baker, Jr, M. B., Rogers, J. J., Clary, W. P., Kovner, J. L., Larson, F. R., Avery, C. C. and Campbell, R. E. 1974, *Opportunities for increasing water yields and other multiple use values on ponderosa pine forest lands,* US Department of Agriculture Forest Service Research Paper

RM–129, Rocky Mountain Forest and Range Experiment Station, Fort Collins, Colorado.

Bruijnzeel, L. A. 1990, *Hydrology of moist tropical forests and effects of conversion: a state of knowledge review,* National Committee of the Netherlands for the International Hydrological Program of United Nations Educational, Scientific and Cultural Organisation; International Institute for Aerospace Survey and Earth Sciences, Programme on Geo-information for Environmentally Sound Management of Tropical Resources; and Inter-national Association of Hydrological Sciences, n.p.

Bureau of Outdoor Recreation 1972, *The 1965 survey of outdoor recreation activities,* US Government Printing Office, Washington, DC.

Bureau of Outdoor Recreation 1973, *Outdoor recreation: a legacy for America,* US Government Printing Office, Washington, DC.

Burgman, M. A., unpublished, *Characteristics and delineation of the old-growth estate.* Draft paper. University of Melbourne, Melbourne.

Burgman, M., Church, R., Ferguson, I., Gijsbers, R., Lau, A., Lindenmayer, D., Loyn, R., McCarthy, M. and Vandenburg, W. 1994, Wildlife planning using FORPLAN: a review and examples from Victorian forests, *Australian Forestry,* 57 (3), 131–40.

Burgman, M. A. and Ferguson, I S. 1995, *Rainforest in Victoria —a review of the scientific basis of current and proposed protection measures.* Department of Conservation and Natural Resources, Forest Service Technical Reports 95–4, Department of Conservation and Natural Resources, Melbourne.

Burgman, M. A., Ferson, S. and Akcakaya, H. R. 1993, *Risk assessment in conservation biology*, Chapman & Hall, London.

Campbell, I. C. and Doeg, T. J. 1985, The impact of timber production and harvesting on aquatic values. In *Report of the Board of Inquiry into the Tmber Industry in Victoria,* Vol. 2. *Commissioned papers* (I. S. Ferguson, chairman), Government Printer, Melbourne.

Campbell, R. G. 1974, *A stochastic analysis of some management decisions for Mountain Ash.*, Ph.D. thesis, University of Melbourne.

Carlson, S. 1965, [1939], *A study on the pure theory of production,* Reprints of Economic Classics, Augustus M. Kelley, New York.

Caterpillar Tractor Company 1988, *Caterpillar performance handbook* (19th edn), Caterpillar Tractor Company, Peoria, Illinois.

Chambers, R. L., Saxena, N. C. and Shah, T. 1989, *To the hands of the poor: water and trees,* Oxford and IBH Publishing, New Delhi.

Cicchetti, C. J. 1973, *Forecasting the demand for recreation,* Lexington Books, Lexington.

Ciriacy-Wantrup, S. V. 1963, *Resource conservation: economics and policies* (rev. edn), University of California Division of Agricultural Sciences, Agricultural Experiment Station, Berkeley.

Clarke, H. R. 1994, Forest rotation and streamflow benefits, *Australian Forestry*, 57(1), 37–44.

Clawson, M. and Knetsch. J. L. 1966, *The economics of outdoor recreation*, Johns Hopkins Press, Baltimore.

Clinnick, P. F. 1985, Buffer strip management in forest operations: a review, *Australian Forestry*, 48 (1), 34–45.

Clinnick, P. F. and Incerti, M. 1985, Compaction effects from forest operations at two sites in Victoria. Internal research report, Land Protection Service, Department of Conservation, Forests and Lands, Victoria, Australia.

Clinnick, P. F., Cummings, D. J. and Howe, D. F. 1984, An approach to estimating slope limits for forest harvesting. Paper presented at the Australian Society of Soil Science, National Soils Conference, Brisbane, Australia, 1984.

Clutter, J. L., Fortson, J. C., Pienaar, L. V., Brister, G. H. and Bailey, R. L. 1983, *Timber management: a quantitative approach*, John Wiley & Sons, New York.

Common, M. and Perrings, C. 1992, Towards an ecological theory of sustainability, *Ecological Economics*, 6 (1), 7–34.

Constanza, R. and Daly, H. E. 1992, Natural capital and sustainable development, *Conservation Biology*, 6 (1), 37–45.

Cornish, P. M. 1981, Modelling soil mosisture in a *Pinus radiata* plantation. Paper presented at XVII International Union of Forest Research Organisations World Congress, Kyoto, Japan, 1981.

Crowe, S. 1966, *Forestry in the landscape*, Forestry Commission Booklet 18, HMSO, London.

Cubbage, F. W., O'Laughlin, J. and Bullock III, C. S. 1993, *Forest resource policy*, John Wiley & Sons, New York.

Cummings, R. G., Cox, Jr, L. A. and Freeman III, M., A. 1984, General methods for benefits assessment. In *Evaluation of the state-of-the art in benefits assessment methods for public policy purposes* (comp. Arthur D. Little Co.), Report to the Division of Policy Research and Analysis, National Science Foundation, Cambridge, Massachusetts.

Cunningham, T. M. 1960, *The natural regeneration of Eucalyptus regnans*. University of Melbourne, School of Forestry Bulletin No. 1, Melbourne University Press, Parkville, Victoria.

Dasgupta, P. S. and Heal, G. 1979, *Economic theory and exhaustible resources*, Cambridge University Press, Cambridge.

Dasgupta, P., Sen, A. and Marglin, S. 1972, *Guidelines for project evaluation*, United Nations Industrial Development Organisation Project Formulation on Evaluation Series No 2, United Nations, New York.

Davis, L. S. and Johnson, K. N. 1987, *Forest management* (3rd edn), McGraw-Hill, New York.

Department of Conservation and Environment, Victoria. 1992, *Forest management plan: Otway Forest Management Area,* Department of Conservation and Environment, Melbourne.

Department of Conservation, Forests and Lands, Victoria. 1988, *Code of Forest Practice for Timber Production,* Department of Conservation, Forests and Lands, Melbourne.

Department of Primary Industries and Energy and Department of Environment, Sport and Territories. 1995, *National Forest Conservation Reserves: Commonwealth proposed criteria,* Department of Primary Industries and Energy and Department of Environment, Sport and Territories, Canberra.

Diamond, P. A. and Hausman, J. A. 1994, Contingent valuation: is some number better than no number, *Journal of Economic Perspectives,* 8 (4), 45–64.

Dickinson, R. E. and Henderson-Sellars, A. 1988, Modelling tropical deforestation: A study of GCM land-surface parameterizations, *Quarterly Journal of the Royal Meteorological Society,* 114, 439–62.

Dixon, J. A. and P. B. Sherman 1991, *Economics of protected areas: a new look at benefits and costs,* Earthscan Publications, London.

Douglass, R. W. 1982, *Forest recreation* (3rd edn), Permagon Press, New York.

Dove, M. 1993, A revisionist view of tropical deforestation and development, *Environmental Conservation,* 20 (1), 17–24.

Dowdle, B. 1965, *Investment theory and forest management planning,* Yale University School of Forestry Bulletin No 67, Yale University, New Haven.

Dykstra, D. P.1984, *Mathematical programming for natural resource management,* McGraw-Hill, New York.

Dyne, G. 1991, *Attributes of old-growth forest in Australia.* Proceedings of a Workshop sponsored by the Natonal Forest Inventory. Bureau of Rural Resources Working Paper No. WP/4/92, Department of Primary Industries and Energy, Canberra.

Ecologically Sustainable Working Groups. 1991, *Ecologically Sustainable Working Groups, Final report —forest use,* AGPS, Canberra.

Elliot, R. and Gare, A. (eds). 1983, *Environmental philosophy,* University of Queensland Press, St Lucia, Queensland.

Faustmann, M., 1995 [1849], On the determination of the value which forest land and immature stands pose for forestry (trans. W. Linnard), *Journal of Forest Economics*, 1 (1), 7–44.

Ferguson, I. S. 1985, *Report of the Board of Inquiry into the Timber Industry in Victoria*, Vol. 1., Government Printer, Melbourne.

Ferguson, I. S. 1995, Timber production and streamflow benefits: a contrary analysis of options, *Australian Forestry* 58 (3):142–6.

Ferguson, I.S. forthcoming, Sustainable management of indigenous public forests: myth, mandala or mandate? *East-West Center Working Document,* in press, East West Center, Honolulu.

Ferguson, I. S. and Bren, L. J. 1991, Forest harvesting and environmental concerns. Paper presented at the Institution of Engineers, Australia, Conference on 'The role of the engineering profession in sustainable development,' Hobart, Tasmania, 1991.

Ferguson, I. S. and Greig, P. J. 1972, What price recreation? *Australian Forestry* , 36 (2), 80–90.

Ferguson, I. S. and Munoz-Reyes Navarro, J. 1992, *Resources needed by producer countries to achieve sustainable management by the year 2000*, Working Document for International Tropical Timber Organisation Expert Panel, International Tropical Timber Organisation, Yokohama.

Ferguson, I. S. and Reilly, J. J. 1976, The social discount rate and opportunity cost of capital in forestry development projects. In *Evaluation of the contribution of forestry to economic development* (ed. A. J. Grayson), pp. 85–93, Forestry Commission of Great Britain, Bulletin No. 56, HMSO, London.

Ferriss, A. L. 1962, *National recreation survey*, Outdoor Recreation Resources Review Commission Study Report 19, US Government Printing Office, Washington, DC.

Fisher, A. C. 1981, *Resource and environmental economics,* Cambridge University Press, Cambridge.

Food and Agriculture Organisation 1991, *1961–1989....2010: wood and wood products*, Food and Agriculture Organisation, Rome.

Food and Agriculture Organisation 1994, *FAO yearbook of forest products, 1981–1992,* Food and Agriculture Organisation, Rome.

Food and Agriculture Organisation and United Nations—Economic Community of Europe 1993, Summary of results for temperate zones (Forest Resource Assessment 1990, In *Proceedings of FAO/ECE meeting of experts on global forest resources assessment in co-operation with UNEP and with the support of FINNIDA (Kotka II)* (ed. N. Nyyssonen), pp. 111–16, Research Papers 469, Finnish Forest Research Institute, Helsinki.

Forest Ecosystem Management Assessment Team 1993, *Forest ecosystem management: an ecological, economic and social assessment,* US Government Printing Office, Washington, DC.

Forestry Commission Tasmania. 1983, *Visual management system: the forest landscape,* Forestry Commission Tasmania Bulletin No.9, Forestry Commission, Hobart.

Forestry Commission Tasmania 1987, *Forest Practices Code,* Forestry Commis-sion, Hobart.

Franklin, J. F., Cromack, K. Jr., Denison, W., McKee, A., Maser, C., Sedell, J., Swanson, F. and Juday, G., 1981, *Ecological attributes of old-growth Douglas-fir forests.* United States Department of Agriculture, Forest Service, General Technical Report, PNW-118, Pacific Northwest Forest and Range Experiment Station. Portland, Oregon.

Gaffney, M. R. 1957, *Concepts of financial maturity of timber and other assets,* Agricultural Information Series No. 62, North Carolina State University, Raleigh.

Gershon, D. 1992, If biological diversity has a price, who sets it and who should benefit? *Nature,* 359, 565.

Glick, D. 1991, Tourism in Greater Yellowstone: maximizing the good, minimizing the bad, eliminating the ugly. In *Nature tourism: managing for the environment* (ed. T. Whelan), pp. 58–74, Island Press, Washington, DC.

Graham, R. L., Turner, M. G. and Dale, V. H. 1990, How increasing CO_2 and climate change affect forests, *Bioscience,* 40, 575–87.

Gregory, G. R. 1955, An economic approach to multiple use, *Forest Science,* 1 (1), 6–13.

Gregory, G. R. 1972, *Forest resource economics,* The Ronald Press Company, New York.

Greig, P. 1977, Data collection for recreation research and planning in Australia, In *Leisure and recreation in Australia* (ed. D. Mercer), pp. 152–72, Sorret Publishing, Malvern, Australia.

Groombridge, B. (ed.). 1992, *Global biodiversity: status of the earth's living resources—a report,* Report compiled by the World Conservation Monitoring Centre, Chapman & Hall, London.

Hall, D. C. and Hall, J. V. 1984, Concepts and measures of natural resource scarcity with a summary of recent trends, *Journal of Environmental Economics and Management,* 11 (4), 363–79.

Hanemann, W. M. 1994, Valuing the environment through contingent valuation, *Journal of Economic Perspectives,* 8 (4), 19–43.

Hansen, J. R. 1978, *Guide to practical project appraisal: social benefit-cost analysis in developing countries*, United Nations International Development Organisation Project Formulation and Evaluation Series No. 3, United Nations, New York.

Harris, L. D. 1984, *The fragmented forest: island biogeography theory and the preservation of biotic diversity*, University of Chicago Press, Chicago.

Hartwick, J. M. 1977, Intergenerational equity and the investing of rents from exhaustible resources, *American Economic Review*, 66, 972–4.

Hartwick, J. M. 1978, Investing returns from depleting renewable natural resource stocks and intergenerational equity, *Economic Letters*, 1, 85–8.

Heritage, Conservation and Recreation Service 1979, *The Third Nationwide Recreation Plan*, US Government Printing Office, Washington, DC.

Hirshleifer, J., DeHaven, J. C. and Milliman, J. W. 1960, *Water supply: economics, technology and policy*, University of Chicago Press, Chicago.

Hof, J. 1993, *Coactive forest management*, Academic Press, San Diego.

Holdgate, M. 1993, Sustainability in the forest, Keynote address to 14th Commonwealth Forestry Conference, *Commonwealth Forestry Review*, 72 (4), 217–25.

Hotelling, H. 1938, The general welfare in relation to problems of taxation and of railway and utility rates, *Econometrica*, 6, 242–69.

Howarth, R. B. 1991, Interemporal equilibria and exhaustible resources: an overlapping generations approach, *Ecological Economics*, 4 (1), 237–52.

Howarth, R. B. and Norgaard, R. B. 1990, Intergenerational property rights, efficiency and social optimality, *Land Economics* , 66 (1), 1–11.

Hyde, W. F. and Newman, D. H. 1992, *Forest economics and policy analysis*, World Bank Discussion Paper 134, World Bank, Washington, DC.

International Tropical Timber Organisation 1991a, *Report of Working Group on ITTO guidelines for the establishment and sustainable management of planted tropical forests*, Paper ITTC(X)/9 Rev. 1, International Tropical Timber Organisation, Yokohama.

International Tropical Timber Organisation 1991b, *Decision 7(XI) Sustainable forest management II*, Paper ITTC(XI)/21, International Tropical Timber Organisation, Yokohama.

International Tropical Timber Organisation 1991c, *Report of Working Group on ITTO guidelines for the establishment and sustainable management of planted tropical forests*, Paper ITTC(X)/9 Rev. 1, International Tropical Timber Organisation, Yokohama.

International Union for Conservation of Nature and Natural Resources 1990, *1990 IUCN red list of threatened animals*, Compiled by the World Conservation Monitoring Centre, International Union for Conservation of Nature and Natural Resources, Gland, Switzerland.

International Union for Conservation of Nature and Natural Resources, United Nations Environment Program and World Wide Fund for Nature 1991, *Caring for the earth: a strategy for sustainable living*, Gland, Switzerland.

Jaeger, J. 1988, *Developing policies for responding to climatic change: a summary of the discussions and recommendations of the workshops held in Villach (28 September-2 October 1987) and Bellagio (9–13 November 1987)*, World Climate Program Impact Studies, WCIP-1, World Meteorological Organisation, Geneva.

Jasisuriya, S. 1992, Economists on sustainability, *Review of Marketing and Agricultural Economics*, 60 (2), 231–41.

Joint Scientific Committee 1990, *Biological conservation of the south-east forests*, Report to the Minister for Resources, Commonwealth of Australia and the Minister for Natural Resources, State of New South Wales, (B. N. Richards, Chairman), AGPS, Canberra.

Kalter, R. J. and Gosse, L. E. 1969, *Outdoor recreation in New York State: projections of demand, economic value, and pricing effects for the period 1970–1985*, Special Cornell Series No. 5, New York State College of Agriculture at Cornell University, Ithaca, NY.

Kanowski, P. J., Savill, P. S., Adlard, P.G., Burley, J., Evans, J., Palmer, J. R. and Wood, P. J. 1992, Plantation forestry. In *Managing the world's forests: looking for balance between conservation and development* (ed. N. P. Sharma), pp. 375–401, Kendall/Hunt Publishing Company, Dubuque, Iowa.

Keating, M. 1993, *The Earth Summit Agenda for Change*, Centre for Our Common Future, Geneva.

Kikeri, S., Nellis, J. and Shirley, M. 1992, *Privatization: the lessons of experience,* The World Bank, Washington, DC.

Kirkpatrick, J. B. and Brown, M. J. 1991, *Reservation analysis of Tasmanian Forests,* Resource Assessment Commission Forest and Timber Inquiry Consultancy Series No. FTC91/16, Resource Assessment Commission, Canberra.

Kohn, R. E. 1993, Measuring the existence value of wildlife: a comment, *Land Economics*, 69 (3), 304–8.

Kuczera, G. 1985, *Prediction of water yield reductions following a bushfire in ash-mixed species eucalypt forest*, Report No. MMBW-W-0014, Melbourne Metropolitan Board of Works, Melbourne.

Leitch, C. J., Flinn, D. W. and van de Graaff, R. H. M. 1983, Erosion and nutrient loss resulting from Ash Wednesday (February 1983) wildfires: a case study, *Australian Forestry*, 46 (3),173–80.

Leonard, M. and Hammond, R. 1984, *Landscape character types of Victoria:*

with frames of reference for scenic quality assessment, Forests Commission, Victoria.

Lesica, P. and Allendorf, F. W. 1991, Are small populations of plants worth preserving? *Conservation Biology,* 5 (2), 135–9.

Leslie, A. J. 1985, *A review of pulpwood royalty policy,* University of Melbourne School of Forestry Bulletin No. 3, University of Melbourne, Melbourne.

Leuschner, W. A. 1984, *Introduction to forest resource management,* John Wiley & Sons, New York.

Lind, R. C. 1990, Re-assessing the Government's discount policy in the light of new theory and data in a world economy with a high degree of capital mobility, *Journal of Environmental Economics and Management,* 18 (2), 8–28.

Lindblom, C. E. 1959, The science of muddling through, *Public Adminstratrion Review,* 19, 79–99.

Lindenmayer, D. B. and Possingham, H. P. 1994, *The risk of extinction: ranking management options for Leadbeater's possum using population viability analysis,* Centre for Resource and Environmental Studies, Australian National University, Canberra.

Lindenmayer, D. B., Thomas, V. C., Lacey, R. C. and Clark, T. W. 1991, *Population viability analysis (PVA): the concept and its applications, with a case study of Leadbeater's possum,* Gymnobelideus leadbeateri McCoy, Resource Assessment Commission Forest and Timber Inquiry Consultancy Series No. FTC91/18, Resource Assessment Commission, Canberra.

Little, I. M. D. and Mirrlees, J. A. 1969, *Manual of industrial project analysis in developing countries,* Vol. 2, *Social cost-benefit analysis,* Development Centre of the Organisation for Economic Co-operation and Development, Paris.

Litton, Jr, R. B. 1968, *Forest landscape description and inventories —a basis for planning and design,* Pacific Southwest Forest and Range Experiment Station Research Paper PSW-49, US Department of Agriculture Forest Service, Berkeley.

Lucas, G. and Synge, H. 1978, *The IUCN plant red data book: comprising red data sheets on 250 selected plants threatened on a world scale,* International Union for the Conservation of Nature and Natural Resources, Morges, Switzerland.

Lucas, R. C. 1990, Wilderness use and users: trends and projections. In *Wilderness management* (eds J. C. Hendee, G. H. Stankey and R. C. Lucas), pp. 355–98, North American Press, Fulcrum Publishing, Golden, Colorado.

MacArthur, R. H. and Wilson, E. O. 1967, *The theory of island bio-geography*, Princeton University Press, Princeton.

Mace, G. M. and Lande, R. 1991, Assessing extinction threats: toward a reevaluation of IUCN threatened species categories, *Conservation Biology* 5 (2), 148–57.

Manning, R. E. 1985, *Studies in outdoor recreation: search and research for satisfaction*, Oregan State University Press, Corvallis, Oregon.

Margules, C. R. and Austin, M. P. 1991, *Nature conservation: cost effective biological surveys and data analysis*, CSIRO, Australia.

Marshall, A. 1966, *Principles of economics: an introductory volume* (8th edn.), Macmillan, London.

McHarg, I. L. 1969, *Design with nature*, Natural History Press, Garden City, New York.

McNeely, J. A. 1988, *Economics and biological diversity: developing and using economic incentives to conserve biological resources.* International Union for the Conservation of Nature and Natural Resources, Gland Switzerland.

McNeely, J. A. (ed.). 1993, *Parks for life: report of the 4th World Congress on National Parks and Protected Areas*, World Conservation Union and World Wide Fund for Nature, Gland, Switzerland.

Mitchell, R. C. and Carson, R. T. 1989, *Using surveys to value public goods: the contingent valuation method*, Resources for the Future, Washington, DC.

Moran, A., Chisholm, A. and Porter, M. (eds). 1991, *Markets, resources and the environment*, Allen & Unwin, North Sydney.

Myers, K., Margules, C. R. and Musto, I. 1983, *Survey methods for nature conservation*, 2 vols, CSIRO Division of Water and Land Resources, Canberra.

Nash, R. 1973, *Wilderness and the American mind*, (2nd edn), Yale University Press, New Haven.

Nash, R. F. 1989, *The rights of nature*, University of Wisconsin Press, Madison.

National Academy of Sciences, National Academy of Engineering and Institute of Medicine 1992, *Policy implications of greenhouse warming: mitigation, adaptation and the science base*, Report by the Panel on Policy Implications of Greenhouse Warming, National Academy Press, Washington, DC.

Noble, I. R. and Slatyer, R. O. 1981, Concepts and models of succession in vascular plant communities subject to recurrent fire. In *Fire and the Australian biota* (eds A. M. Gill, R. H. Groves and I. R. Noble), pp. 311–35, Australian Academy of Science, Canberra.

Nordhaus, W. D. 1973, The allocation of energy resources, *Brookings Papers on Economic Activity*, 529–70.

Norgaard, R. B. 1992, *Sustainability and the economics of assuring assets for future generations*, Office of the Regional Vice President Working Papers WPS 832, World Bank, Asia Regional Development Office, n.p.

Norgaard, R. B. and Howarth, R. B. 1992, Environmental valuation under sustainable development, *American Economic Review*, 82 (2), 473–7.

O'Hara, A. J., Faaland, B. H. and Bare, B. B. 1989, Spatially constrained timber harvest scheduling, *Canadian Journal of Forest Research*, 19, 715–24.

O'Regan, M. and Bhati, U. N. 1991, *Pricing and allocation of logs in Australia,* Australian Bureau of Agricultural and Resource Economics Discussion Paper 91.7, Australian Bureau of Agricultural and Resource Economics, Canberra.

O'Regan, M. and Bhati, U. N. 1990, *Fiscal approaches of state forest management agencies,* Resource Assessment Commission, Forest and Timber Industry Inquiry Consultancy Series No. FTC91/01, Australian Bureau of Agriculture and Resource Economics, Canberra.

O'Shaughnessy. P. J. and Jayasuriya, M. D. A. 1991, Managing the ash-type forests for water protection in Victoria, In *Forest management in Australia* (eds F.H. McKinnell, E. R. Hopkins and J. E. D. Fox), pp. 301–63, Surrey Beatty & Sons, Sydney.

Pearce, D. W. and Turner, R. K. 1990, *Economics of natural resources and the environment,* Johns Hopkins Press, Baltimore.

Pearce, D. W. and Warford, J. J. 1993, *World without end: Economics, environment and sustainable development*, Oxford University Press, Oxford.

Persson, R. 1974, *World forest resources: a review of the worlds forest resources in the early 1970s*, Skogsh & Ogskolan, Stockholm.

Poore, D., Burgess, P., Palmer, J., Rietbergen, S. and Synnott, T. 1989, *No timber without trees: sustainability in the tropical forest,* Earthscan Publications, London.

Portney, P. R. 1994, The contingent valuation debate: why economists should care, *Journal of Economic Perspectives,* 8 (4), 3–17.

Pyne, S. J. 1991, *Burning bush: a fire history of Australia,* Allen & Unwin, North Sydney.

Raison, J., Geary, P., Kerruish, B., Connell, M., Jacobsen, K., Roberts, E. and Hescock, R. 1995, Opportunities for more productive management of regrowth eucalypt forests: A case study in the *E. sieberi* forests of East Gippsland. In *Applications of new technologies in forestry* (eds L. Bren and C. Greenwood), Proceedings of the Institute of Foresters of Australia

16th Biennial Conference, pp. 31–7, The Institute of Foresters of Australia, Canberra.

Read, Sturgess and Associates 1992, *Evaluation of the economic values of wood and water for the Thomson Catchment,* Consultancy report prepared for Department of Conservation and Natural Resources, Victoria and Melbourne Water, Victoria, Read Sturgess and Associates, Kew.

Read, Sturgess and Associates and Tasman Economic Research 1994, *Phase two of the study into the economic evaluation of wood and water for the Thomson Catchment,* Consultancy report prepared for Department of Conservation and Natural Resources, Victoria and Melbourne Water, Victoria, Read, Sturgess and Associates, Kew.

Reid, L. M. and Dunne, T. 1984, Sediment production from road surfaces, *Water Resources Research,* 20 (11), 1753–61.

Repetto, R. and Gillis, M. (eds). 1988, *Public policies and the misuse of forest resources,* Cambridge University Press, Cambridge.

Resource Assessment Commission 1991, *Commentaries on the Resource Assessment Commission's contingent valuation survey of Kakadu Conservation Zone,* Resource Assessment Commission, Canberra.

Resource Assessment Commission 1992, Estimating the preservation value of forests, In *Forest and Timber Inquiry Final Report,* Vol. 2B, AGPS, Canberra.

Richards, M. 1993, The potential of non-timber forest products in sustainable natural forest management in Amazonia, *Commonwealth Forestry Review,* 72 (1), 21–7.

Rothenburg, J. 1993, Economic perspectives on time comparisons; alternative approaches to time discounting. In *Global accord: environmental challenges and international responses* (ed. N. Choucri), pp. 355–97, The MIT Press, Cambridge, Massuchusetts.

Rothschild, Lord, 1971, *The organization and management of government research and development,* HMSO, London.

Ruitenbeek, J. H. 1990, *Economic analysis of tropical forest conservation initiatives: examples from West Africa,* World Wildlife Fund for Nature, Godalming, Surrey, UK.

Ruitenbeek, J. H. 1992, The rainforest supply price: a tool for evaluating rainforest conservation expenditures, *Ecological Economics,* 6, 57–78.

Samuelson, P. A. 1995 [1976], Economics of forestry in an evolving society, *Journal of Forest Economics,* 1 (1), 115–49.

Sandler, T. 1993, Tropical deforestation: Markets and market failures, *Land Economics,* 69 (3), 225–33.

Satterlund, D. R. and Adams, P. W. 1992, *Wildland watershed management,* (2nd edn), John Wiley & Sons, New York.

Sedjo, R. A. 1989, Forests to offset the greenhouse effect, *Journal of Forestry,* 87, 12–15.

Shaffer, M. 1987, Minimum viable populations: coping with uncertainty. In *Viable populations for conservation* (ed. M. E. Soulé), pp. 69–86, Cambridge University Press, Cambridge.

Shaw, G. and Williams, A. M. 1994, *Critical issues in tourism: a geographical perspective,* Blackwell, Oxford.

Silviconsult 1990, *Natural forest rehabilitation study,* 2 vols, Report to Asian Development Bank, Silviconsult, n.p.

Sinden, J. A. 1974, A utility approach to the valuation of recreational and aesthetic experiences, *American Journal of Agricultural Economics,* 56 (1), 61–72.

Sinden, J. A. and Worrell, A. C. 1979, *Unpriced values: decisions without market prices,* John Wiley & Sons, New York.

Slade, M. E. 1982, Trends in natural resource commodity prices: an analysis of the time domain, *Journal of Environmental Economics and Management* , 9, 22–37.

Solow, R. 1992, *An almost practicable step towards sustainability,* Invited lecture on the occasion of the Fortieth Anniversary of Resources for the Future, Resources for the Future, Washington DC.

Stadtmuller, T. 1987, *Cloud forests in the humid tropics: a bibliographic review,* The United Nations University, Tokyo and Centro Agronomico Tropical de Investigacion y Ensenanza, Turrialba, Costa Rica.

Stankey, G. H. 1973, *Visitor perceptions of wilderness recreation carrying capacity,* Research Paper INT-142, US Department of Agriculture Forest Service, Intermountain Forest and Range Experiment Station, Ogden, Utah.

Stevens, T. H., More, T. A. and Glass, R. J. 1993, Measuring the existence value of wildlife: Reply, *Land Economics* , 69 (3), 309–12.

Stiglitz, J. E. 1986, *Economics of the public sector,* (2nd edn), W.W. Norton & Company, New York.

Stroup, R. L. and Baden, J. A. 1983, *Natural resources: bureaucratic myths and environmental management,* Ballinger Publishing Company, Cambridge, Massachusetts.

Tietenberg, T. H. 1992, *Enviromental and natural resource economics,* (3rd edn), Harper Collins Publishers, New York.

US Department of Agriculture, Forest Service 1973, *National forest landscape management,* Vol. 1, Agriculture Handbook No. 434, US Department of Agriculture, Forest Service, Washington, DC.

US Department of Agriculture, Forest Service 1974, Chapter 1—The visual management system. In *National forests landscape management,* Vol. 2,

Agriculture Handbook No. 462, US Department of Agriculture Forest Service, Washington.

United Nations 1984, *International Tropical Timber Agreement, 1983*, United Nations, New York.

United Nations Educational, Scientific and Cultural Organisation 1987, *Operational guidelines for the implementation of the World Heritage Convention*, Prepared by the Inter-Governmental Committee for the Protection of World Cultural and Natural Heritage, United Nations Educational, Scientific and Cultural Organisation, Paris.

Vanclay, J. 1994, *Modelling forest growth and yield: applications to mixed tropical forests*, Centre for Agriculture and Biosciences International, Wallingford, UK.

Whitmore, T. C. 1990, *An introduction to tropical rain forests*, Clarendon Press, Oxford.

Wischmeier, W. H. and Smith, D. D. 1978, *Predicting rainfall erosion losses: a guide to soil conservation*, US Department of Agriculture Handbook 537, US Government Printing Office, Washington, DC.

Woodwell, G. M. 1992, The role of forests in climate change. In *Managing the world's forests: looking for balance between conservation and development* (ed. N. P. Sharma), pp. 75–91, Kendall/Hunt Publishing Company, Dubuque, Iowa.

World Commission on Environment and Development 1987, *Our common future*, Oxford University Press, Oxford.

Index